Are You Smart Enough to Work at Google?

"A neat little manifesto on interview technique...Touring through a huge number of puzzles, [Poundstone] provides a truly exhaustive account... Tackling [them] is incredibly gratifying."

New Scientist

"Poundstone... display[s] his scientific knowledge, mathematical fluency and knack for explaining the arcane in playfully precise sentences."

Bloomberg Businessweek

"Serious ammunition to pack for your next job interview."

Kirkus Reviews

"A great book." *Business Life*

"Poundstone's energetic, compelling writing... makes the book fun even for non-job-seekers."

Publishers Weekly

"Splendidly exhaustive."

International Herald Tribune

Also by William Poundstone

Big Secrets

The Recursive Universe

Bigger Secrets

Labyrinths of Reason

The Ultimate

Prisoner's Dilemma

Biggest Secrets

Carl Sagan: A Life in the Cosmos

How Would You Move Mount Fuji?

Fortune's Formula

Gaming the Vote

Priceless

Are You Smart Enough to Work at Google?

Fiendish Puzzles and
Impossible Interview Questions
from the World's Top Companies

William Poundstone

ONEWORLD

A Oneworld Book

First published in the United Kingdom and the Commonwealth by
Oneworld Publications in 2012
This paperback edition published by Oneworld Publications 2013

Originally published in the United States of America by Little, Brown and
Company, a division of the Hachette Book Group, Inc., New York.

A CIP record for this title is available from
the British Library

ISBN 978-1-85168-955-2
eISBN 978-1-78074-075-1

Printed and bound in Great Britain by TJ International Ltd

Oneworld Publications
10 Bloomsbury Street, London WC1B 3SR, England

To the memory of Martin Gardner (1914–2010),
whose influence on this book's subject matter is considerable.

A hundred prisoners are each locked in a room with three pirates, one of whom will walk the plank in the morning. Each prisoner has ten bottles of wine, one of which has been poisoned; and each pirate has twelve coins, one of which is counterfeit and weighs either more or less than a genuine coin. In the room is a single switch, which the prisoner may either leave as it is or flip. Before being led into the rooms, the prisoners are all made to wear either a red hat or a blue hat; they can see all the other prisoners' hats but not their own. Meanwhile, a six-digit prime number of monkeys multiply until their digits reverse, then all have to get across a river using a canoe that can hold at most two monkeys at a time. But half the monkeys always lie and the other half always tell the truth. Given that the Nth prisoner knows that one of the monkeys doesn't know that a pirate doesn't know the product of two numbers between 1 and 100 without knowing that the $N + 1$th prisoner has flipped the switch in his room or not after having determined which bottle of wine was poisoned and what color his hat is, what is the solution to this puzzle?

—*Internet parody of a job interview question*

Contents

Are You Smart Enough to Work at Google?

One

Outnumbered at the Googleplex

What It Takes to Get Hired at a Hyperselective Company

Jim was sitting in the lobby of Google's Building 44, Mountain View, California, surrounded by half a dozen others in various states of stupor. All were staring dumbly at the stupidest, most addictive TV show ever. It is Google's live search board, the ever-scrolling list of the search terms people are Googling at this very instant. Watching the board is like picking the lock to the world's diary, then wishing you hadn't. For one moment, the private desires and anxieties of someone in New Orleans or Hyderabad or Edinburgh are broadcast to a select audience of voyeurs in Google lobbies—most of them twenty- and thirty-year-olds awaiting a job interview.

giant-print Bibles
overseeding
Tales of Phantasia
world's largest glacier
JavaScript
man makeup
purpose of education
Russian laws relating to archery

Jim knew the odds were stacked against him. Google was receiving a million job applications a year. It was estimated that only about 1 in 130 applications resulted in a job. By comparison, about 1 in 14 students applying to Harvard University gets accepted. As at Harvard, Google employees must overcome some tall hurdles.

Jim's first interviewer was late and sweaty: he had biked to work. He started with some polite questions about Jim's work history. Jim eagerly explained his short career. The interviewer didn't look at him. He was tapping away at his laptop, taking notes.

"The next question I'm going to ask," he said, "is a little unusual.

? You are shrunk to the height of a penny and thrown into a blender. Your mass is reduced so that your density is the same as usual. The blades start moving in sixty seconds. What do you do?"*

The interviewer had looked up from his laptop and was grinning like a maniac with a new toy.

"I would take the change in my pocket and throw it into the blender motor to jam it," Jim said.

The interviewer's tapping resumed. "The inside of a blender is sealed," he countered, with the air of someone who had heard it all before. "If you could throw pocket change into the mechanism, then your smoothie would leak into it."

"Right... um... I would take off my belt and shirt, then. I'd tear the shirt into strips to make a rope, with the belt, too, maybe. Then I'd tie my shoes to the end of the rope and use it like a lasso..."

---

* Whenever you see the **?** symbol in this book, it means there's a discussion in the answer section, starting on p. 137.

Furious key clicks.

"I don't mean a lasso," Jim plowed on. "What are those things Argentinean cowboys throw? It's like a weight at the end of a rope."

No answer. Jim now felt his idea was lame, yet he was compelled to complete it. "I'd throw the weights over the top of the blender jar. Then I'd climb out."

"The 'weights' are just your shoes," the interviewer said. "How would they support your body's weight? You weigh more than your shoes do."

Jim didn't know. That wasn't the end of it. The interviewer had suddenly warmed to the topic. He began ticking off quibbles one by one. He wasn't sure whether Jim's shirt—shrunken with the rest of him—could be made into a rope that would be long enough to reach over the lip of a blender. Once Jim got to the top of the jar—*if* he got there—how would he get down again? Could he realistically make a rope in sixty seconds?

Jim didn't see where a word like *realistic* came into play. It was as if Google had a shrinking ray and was planning to try it out next week.

"It was nice meeting you," the interviewer said, extending a still-damp hand.

We live in an age of desperation. Never in living memory has the competition for job openings been more intense. Never have job interviews been tougher. This is the bitter fruit of the jobless recovery and the changing nature of work.

For some job seekers, Google is the shining city on the hill. It's where the smartest people do the coolest things. In the U.S., Google regularly ranks at or near the top of *Fortune* magazine's list of "100 Best Companies to Work For." The Google Mountain View campus (the "Googleplex") is a cornucopia of amenities for its presumably lucky employees. There are eleven gourmet restaurants serving free, organic, locally grown food; climbing walls and pools for swimming in place; mural-size whiteboards for sharing spontaneous thoughts; Ping-Pong, table football, and air-hockey tables; cutesy touches like red phone booths and topiary dinosaurs. Google employees have access to coin-free laundry machines, free flu shots, foreign language lessons, car washes, and oil changes. There is shuttle service between home and work; $5,000 rebates for buying a hybrid; communal scooters for anyone's use on campus. New parents get $500 for takeaway meals and eighteen weeks' leave to bond with their infant. Google pays the income tax on health benefits for same-sex domestic partners. All employees get an annual ski trip. The perks aren't necessarily about generosity, and unlike the workplace gains of previous generations, they haven't been negotiated by unions or individuals. It's good business for Google to offer such benefits in an industry so dependent on attracting the top talent. The benefits not only keep employees happy but also keep everyone else with their noses pressed against the glass.

Google is not so exceptional as you might think. Today's army of unemployed has made every company a Google. Unsexy firms now find themselves with multiple well-qualified applicants for each position. That is very good for the companies that are able to hire. Like Google, they get to cherry-pick the top talent in their fields. It's not so good for the applicants. They are confronting harder, ruder, more invasive vetting than ever before.

This is most evident in the interviews. There are, of course, many types of questions traditionally asked in job interviews. These include the "behavioral" questions that have almost become clichés:

> "Tell me about a situation where you just couldn't get along with a team member."
>
> "Describe a time when you had to deal with a rude customer."

"What is your biggest failure in life?"
"Did you ever find yourself unable to meet a deadline? What did you do?"
"Describe the most diverse team you ever managed."

There are questions relating to business:

"How would you describe Holland & Barrett to a person visiting from another country?"
"Tell me how Waitrose competes with Tesco, and how we should reposition our brand to gain market share."
"How would you get more customers for Halifax Bank?"
"What challenges will Starbucks face in the next ten years?"
"How would you monetize Facebook?"

Then there's work sampling. Rather than asking job candidates what they can do, companies expect them to demonstrate it within the interview. Sales managers have to devise a marketing plan. Attorneys draft a contract. Software engineers write code.

Finally, there are open-ended mental challenges—something for which Google is particularly known. Questions like "thrown into a blender" are an attempt to measure mental flexibility and even entrepreneurial potential. That's been important at Google because of the company's fast growth. A person hired for one job may be doing something else in a few years. Work sampling, while valuable, tests only a particular set of skills. The more offbeat questions attempt to gauge something that every company wants but few know how to measure: the ability to innovate.

For that reason, many of Google's interview questions have spread to companies far beyond Mountain View. Google's "brand" is now estimated to be the most valuable in the world, worth $86 billion, according to Millward Brown Optimor. Success breeds imitation. Corporate types vow to "be more like Google" (whatever that means for the kitchen flooring industry). Not surprisingly, that includes hiring.

What Number Comes Next?

The style of interviewing at Google is indebted to an older tradition of using logic puzzles to test job candidates at technology companies. Consider this one. The interviewer writes six numbers on the room's whiteboard:

10, 9, 60, 90, 70, 66

The question is, what number comes next in the series?

Similar riddles have been used on psychological tests of creativity. Most of the time, the job applicant stumbles around, gamely trying to make sense of a series that gives every indication of being completely senseless. The majority of candidates give up. A lucky few have a flash of insight.

Forget maths. Spell out the numbers in plain English, which gives you the following:

ten
nine
sixty
ninety
seventy
sixty-six

The numbers are in order of how many letters are in their names!

Now look more closely. Ten is not the only number you can spell with three letters. There's also one, two, and six. Nine is not the only four-letter number; there's zero, four, and five. This is a list of the largest numbers that can be spelled in a given number of letters.

Now for the payoff, *what number comes next?* Whatever number follows sixty-six should have nine letters in it (not counting a possible hyphen) and should be the *largest* nine-letter number. Play around with it, and you'll probably come up with ninety-six. It doesn't look like you can get anything above 100 because that would start "one hundred," requiring ten letters and up.

You might wonder why the list doesn't have 100 ("hundred") in place of 70 ("seventy"). "Million" and "billion" have seven letters, too. A reasonable guess is they're using cardinal numbers spelled in correct stylebook English. The way you write out the number 100 is "one hundred."

In the *On-Line Encyclopedia of Integer Sequences,* you can type in a series of numbers and it tells you what numbers come next. You're *not* allowed to use it with this interview question, of course, but the website's answer for this sequence is 96. In recent years, companies in all sorts of industries have adopted this question for interviews. Often the interviewer throws it in just to make the poor candidate squirm. At many of these companies, the one and only correct answer is 96.

Not at Google. In Mountain View, 96 is considered to be an acceptable answer. A better response is

10,000,000,000,000,000,000,000,000,000,000,000,000,000,
000,000,000,000,000,000,000,000,000,000,000,000,000,000,
000,000,000,000,000,000

A.k.a. "one googol."

That's not the best answer, though. The preferred response is

100,000,000,000,000,000,000,000,000,000,000,000,000,000,
000,000,000,000,000,000,000,000,000,000,000,000,000,000,
000,000,000,000,000,000
Ten googol.

That response can be traced back to 1938 or thereabouts. Nine-year-old Milton Sirotta and his brother Edwin were taking a stroll one day with their uncle in the New Jersey Palisades. The uncle was Edward Kasner, a Columbia University mathematician already somewhat famous as the first Jew to gain tenure in the sciences at that Ivy League institution. Kasner entertained the boys by talking about a topic calculated to appeal to bookish nine-year-olds, namely the number that could be written as a "1" followed by a hundred zeros. Kasner challenged his nephews to invent a name for the number. Milton's suggestion was "googol."

That word appeared in the 1940 book that Kasner wrote with James Newman, *Mathematics and the Imagination.* So did the name for an even bigger number, the "googolplex," defined as 10 raised to the power of a googol. Both words caught on and have permeated pop culture, turning up on *The Simpsons*—and as the name for the search engine devised by Larry Page and Sergey Brin. According to Stanford's David Koller,

> Sean [Anderson] and Larry [Page] were in their office, using the whiteboard, trying to think up a good name—something that related to the indexing of an immense amount of data. Sean verbally suggested the word "googolplex," and Larry responded verbally with the shortened form, "googol" (both words refer to specific large numbers). Sean was seated at his computer terminal, so he executed a search of the Internet domain name registry database to see if the newly

suggested name was still available for registration and use. Sean is not an infallible speller, and he made the mistake of searching for the name spelled as "google.com," which he found to be available. Larry liked the name, and within hours he took the step of registering the name "google.com" for himself and Sergey.

Edward Kasner died in 1955 and never saw his number's namesake. More recently, the googol-Google lineage has become a touchy issue. In 2004, Kasner's great-niece, Peri Fleisher, complained that Page and Brin's company had appropriated the word without compensation. Fleisher said she was exploring her legal options. (The best headline ran, "Have Your Google People Talk to My 'Googol' People.")

The googol-Google puzzle has layers like an onion. First you have to realize that the spelling of the numbers, rather than their mathematical properties, is relevant. That's hard enough. Then you have to know about, and remember, Kasner's number. An average mortal would think himself clever to come up with "one googol" and be ready to call it a day. But there's still the final layer. "Ten googol" is bigger than "one googol" and ought to be the answer.

Imagination and Invention

Is this question too insanely hard to ask of a job candidate? Not at Google. But puzzles like this have drawbacks as interview questions. The answer here is a simple matter of insight: either you get it or you don't. There isn't a process of deduction to relate, and thus there is no way to distinguish someone who solves the problem from someone who already knew the answer. At Google, of all places, anyone applying for a job knows how to use a search engine. It's expected that candidates will Google for advice on

Google interviews, including the questions asked. Consequently, Google encourages its interviewers to use a different type of question, more open ended, with no definitive "right answer." In the Google philosophy, good interview questions are like take-home tests. The challenge is to come up with an answer the interviewer has never heard before that's *better* than any answer he's heard.

Google's interviewers "are not warm and fuzzy people," as one applicant told me. Another word you hear a lot is "numb"—the utter lack of emotional affect. The interviewer sits, blandly tapping at a laptop. You say something you think is brilliant... no reaction. The keystroke rate doesn't change.

This is by design. Google's mental challenges tend to be cryptic. Candidates are not to be told whether their train of thought is getting "colder" or "warmer," or whether their ultimate answer is right or wrong. Google's challenges often have more than one good answer. Some are considered good, some are banal, and some are brilliant. The interviewee can leave the room with little idea how well he or she did in the interview. This has led to intense speculation and outright paranoia among Google candidates. It has also led to the curious phenomenon of other companies' adopting Google's interview questions without really knowing what the answer is supposed to be.

The quintessential Google perk isn't sashimi or massages. It's the 20 percent project. Google engineers are allowed to spend one day a week on a project of their choosing. That's a fantastic gamble. You can't easily imagine Procter and Gamble giving its staff a day a week to dream up new shampoos. At Google, it works. It's been reported that over half of Google's revenue now comes from ideas that began as 20 percent time projects. The list includes Gmail, Google Maps, Google News, Google Sky, and Google Voice.

How do you measure a talent for invention? Business schools have been asking that question for decades. It's clear that many intelligent people don't have that extra spark, whatever it is. The issue was put well by Nikolay Gogol (whose name is a frequent misspelling for "googol" and "Google"). In his story "The Overcoat," Gogol remarks on "the abyss that separates tailors who only put in linings and do repairs from those who sew new things." Google is betting 20 percent of its engineering-labor costs that it can distinguish the competent software tailors from those capable of creating killer apps out of whole cloth.

The blender riddle encapsulates the process of inventing a new product. You begin by brainstorming. There are many possible answers, and you shouldn't be in a hurry to settle for the first idea that seems "good enough." Coming up with a superior response requires listening carefully to the question's wording. "Imagination is more important than knowledge," Einstein said. You don't have to be an Einstein to answer the question well, but you do need the imagination to connect it to some knowledge you acquired long ago.

For many of us, the knee-jerk response is a facetious one. (One try, posted on a blog: "One might assume that since the blender is about to be turned on, that food will soon be entering, so I'd probably just put my neck to the blade rather than be suffocated by some raunchy health drink.") The two most popular serious answers seem to be (1) lie down, below the blades, and (2) stand to the side of the blades. There ought to be at least a penny's width of clearance between the whirring blades and the bottom or sides of the blender jar.

Another common reply is (3) climb atop the blades and position your center of gravity over the axis. Hold tight. The net centrifugal force will be near zero, allowing you to hold on.

Like many of Google's interview questions, this one leaves a lot unsaid. Who or what has thrown you into the blender, and for what reason? If a hostile being is bent on making a human smoothie, your long-term chance of survival will be small, no matter what you do. Will liquid be added to the blender? Is there a top on it? How long will the blades be spinning? Should the blades spin a long time, answer 3 would make you dizzy. That could cause you to lose consciousness and fall off.

You're welcome to question the interviewer on these points. The canonical responses are "Don't worry about hostile beings," "No liquid will be added," "There's no lid," and "Figure the blades will keep spinning until you're dead."

Another approach is to (4) climb out of the jar. The interviewer will ask how you propose to accomplish that. You don't have suction cups. One bright response is, at that size you're like a fly and can climb glass.

A dumb answer is to (5) use your phone to call or text for help. This depends on your phone's having been shrunk with you and being able to access the nearest (*not* shrunken) mobile phone tower. It also depends on 999 or your Twitter posse sending help in less than sixty seconds.

Still another popular answer is to (6) rip or unravel your clothes to make a "rope" and use it to climb out of the jar. Or (7) use your clothes and personal effects to jam the blades or motor somehow. As we've seen, both have problems.

Of Mice and Men

None of the above answers scores you many points at Google. Current and ex–Google interviewers have told me that the best answer they've heard is (8) *jump out of the jar.*

Huh? The question supplies an important clue, that word *density.* "Being shrunk to the size of a penny" is not a realistic predicament. For starters, it might mean eliminating 99.99+ percent of the neurons in your brain. To deal with a question like this, you have to decide where to suspend disbelief and what to take in earnest. The fact that the interviewer mentions a detail like density is a nudge. It says that things like mass and volume matter in this question (while neuron count might not) and that a successful answer can use simple physics.

In short, the question wants you to consider change-of-scale effects. You probably remember hearing about them in secondary school. An ant is able to lift about fifty times its body weight. It's not because ant muscles are better than human muscles. It's just because ants are *small*. The weight of an ant (or of anything) is in proportion to the cube of its height. The strength of muscles—and the bones or exoskeleton supporting them—depends on their cross-sectional area, which is proportional to the square of height. Were you shrunk to 1/10 your present height, your muscles would be only 1/100 as powerful…but you'd weigh a mere 1/1,000 as much. All else being equal, small creatures are "stronger" in lifting their bodies against gravity. They are more able to bench-press multiples of their own weight.

A classic treatment of change of scale is J. B. S. Haldane's 1926 essay "On Being the Right Size," which you can find by Googling. By using a few basic principles, Haldane was able to account for many mysteries of the biological world. There are no mice or lizards or other small animals in the polar regions. Yet polar bears and walruses thrive. The reason is that small creatures would quickly freeze to death, having a relatively large surface area for their volume. Insects fly easily, but angels are impossible: wings would require too much energy to support a human body.

Haldane's reasoning has been disregarded by decades of cheesy sci-fi movies. Gravity would crush a giant mutant insect like a bug. The advantage would go to the heroes of bad movies like *Honey, I Shrunk the Kids* or *The Incredible Shrinking Man*. Shrunken humans would be fantastically strong, relatively speaking. In the 1957 film, the Incredible Shrinking Man fights off a spider with a needle, lugging it like it's a telephone pole. Actually, he'd be able to maneuver that needle easily.

Do you see where this is going? Were you shrunk to penny size, you'd be strong enough to leap like Superman, right out of the blender.

That is the kernel of a good answer to this question. But Google's interviewers are not just looking for someone who has the basic idea. The best responses supply a coherent argument.

In the mid-1600s Giovanni Alfonso Borelli, a contemporary of Galileo's, deduced this remarkable rule: *everything that jumps, jumps about the same height.* Think about it. Unless you're disabled, you can probably leap about thirty inches, give or take. That's how far you raise your center of gravity. The thirty-inch figure isn't far off the mark for a horse, a rabbit, a frog, a grasshopper, or a flea.

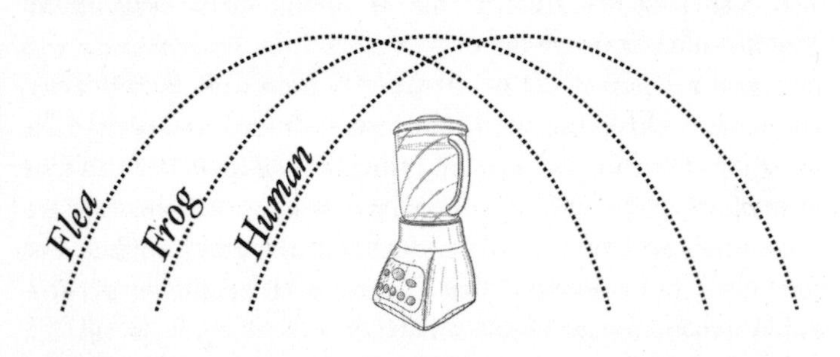

Now, sure, there's variation. A species whose very survival depends on leaping will be optimized for it and do better than one that has little cause to jump. There are species that don't jump at all, like snails, turtles, and elephants. But when you consider the huge variations in size and anatomy, it's amazing that Kobe Bryant and a flea can each put about the same amount of air beneath their feet.

Google doesn't expect anyone to know who Borelli is, but they are impressed with candidates who can replicate his reasoning. This isn't so hard, really. Muscle energy ultimately comes from chemicals—the glucose and oxygen circulating in the blood, and the adenosine triphosphate (ATP) in the muscle cells. The amount of any of these chemicals will be proportional to your body's volume. So if you're shrunk to $1/n$ your usual size, your muscle energy will be reduced by a factor of n^3.

Fortunately, your mass will also be smaller, by the exact same n^3 factor. Consequently, being the size of a penny should neither increase nor decrease how high you can jump (ignoring air resistance). The jar of a blender is about a foot high. As long as you can jump that high right now, you're good.

You might worry about the coming-down part. A blender is something like twenty times the height of a penny. You wouldn't want to fall from twenty times your height now. That shouldn't be a concern, postshrinking. You'll have $1/n^2$ the surface area, versus $1/n^3$ the mass. That means you'll have n times more surface area per mass to resist the fall—and, uh, to hold your guts in when you land. Basically, anything mouse-size or smaller doesn't have to worry about falling from any height. Haldane drew this nice little word picture: "You can drop a mouse down a thousand-yard mine shaft; and, on arriving at the bottom, it gets a slight shock and walks away, provided that the ground is fairly soft. A rat is killed, a man broken, a horse splashes."

Above, I gave an answer (4) in which you simply climb out of the jar like a fly. This, too, can be justified with the change-of-scale argument. You may not think of your hands as sticky, but neither are the footpads of an insect walking up a pane of glass. Try rubbing a pane of glass with your hand: there's resistance. The fact is, every surface clings a little to every other surface. Once you've been shrunk, there will be n times more hand and foot surface per mass and that much more relative clinginess. That might be enough to play Spider-Man.

The Spider-Man answer is still not considered as good as the

Superman answer. Climbing is slow. In proportionate terms, scaling the inside of a twelve-inch-high blender jar would be like an expert rock climber ascending a five-hundred-foot wall. It would be necessary to place each hand or foot carefully. That's going to take time, more than sixty seconds. The blades will be whirring before Spider-Man is over the top. One slip could be fatal. The Superman solution is faster and safer. Should you fail to jump clear of the jar, you'll have another shot at it, probably several.

Scaling Up

As I write these words, 2.5 million British people are out of work. Many of the jobs the unemployed once held are never coming back. People in fields like advertising, retailing, sales, media, and journalism are likely to find themselves interviewing at what they may think of as "technology" companies—only they're not, they're the future of business. This brings them into contact with a new and alien culture of intense interviewing practices.

The blender question is a metaphor. The growth of a company, or of anything we humans care about, is all about change of scale. Solutions that work when something is small do not necessarily work as its scope expands. "For the last year my biggest worry was scaling the business," said Eric Schmidt, then Google CEO, in 2007. "The problem is we're growing so quickly. When you bring people in so quickly there's always the possibility you'll lose the formula."

Difficult interview questions are one way that Google attempts to preserve its formula. Google knows more about "scaling up" than most organizations because of the unique nature of its business and its quick growth. But its experience has lessons for all of us in this slippery, ever-changing, ever-contextual new world. That includes both employers and job seekers.

Hiring at today's selective companies is predicated on the disappointment of the many. This is often a profitable strategy for employers—and it demands a new strategy from job seekers. This book will survey today's supertough interview questions—what they are, how they came to be, and how best to answer them. Whether you're in the job market or not, here's a chance to match wits with employees of some of the world's smartest, most innovative companies. (The questions are a lot of fun, as long as you're not in the hot seat.) Along the way, you'll learn something about the still-profound mystery of creative thinking. Employers will learn much about what works, and what doesn't, in interviews, and why Google's approach—which goes far beyond tough questions—has been so influential. For job seekers, this book will help you avoid being sidelined by a few tricky questions. Often, all it takes to succeed is one good mental leap.

QUESTIONS

A Sample of Today's Interview Interrogations

Try out these questions, popular in job interviews in a wide variety of industries. Answers begin on p. 138.

? When there's a wind blowing, does a round-trip by plane take more time, less time, or the same time?

? What comes next in the following series?

SSS, SCC, C, SC

? You and your neighbor are holding garage sales on the same day. Both of you plan to sell the exact same item. You plan to put your item on sale for £100. The neighbor has informed you that he's going to put his on sale for £40. The items are in identical condition. What do you do, assuming you're not on especially friendly terms with this neighbor?

? You put a glass of water on a record turntable and begin increasing the speed slowly. What will happen first: will the glass slide off, will it tip over, or will the water splash out?

Two

The Cult of Creativity

A History of Human Resources, or Why Interviewers Go Rogue

"You're in an 8-by-8 stone corridor," announced the interviewer. "The Prince of Darkness appears before you."

So begins the tale of a very strange job interview related by Microsoft program manager Chris Sells. "You mean, like, the devil?" asked the unlucky applicant.

"Any prince of darkness will do," she answered. "What do you do?"

"Can I run?"

"Do you want to run?"

"Hmmm. I guess not. Do I have a weapon?"

"What kind of weapon do you want?"

"Um, something with range?"

"Like what?"

"A crossbow?"

"What kind of ammo do you have?"

"Ice arrows?"

"Why?"

"Because the Prince of Darkness is a creature made of fire?"

She liked that. "So what do you do next?"

"I shoot him?"

"No, *what do you do?*" Silence. "*You waste him! You* WASTE *the Prince of Darkness!*"

By this point, the applicant had a question of his own: "Holy crap, what have I gotten myself into?"

He had gotten himself into a not entirely atypical interview in the new economy. In many industries, offbeat interview questions are a badge of coolness. They show how "creative" the workforce is. These questions are a feature of companies where non-HR employees do the interviewing. In highly specialized and creative fields particularly, it's thought that employees better know what questions to ask than human resources people do. This sounds fine in theory. In practice, a largish minority of citizen-interviewers take this as license to go rogue. They ask any offbeat question that pops into their head—and any question they've heard that someone else is asking. Yet it's hard to imagine that questions like the above have much value in selecting people to hire. How did such things come to be?

The deep, dark secret of human resources is that job interviews don't work. This isn't exactly breaking news. Back in 1963, the behavioral scientists Marvin D. Dunnette and Bernard M. Bass wrote,

> The personnel interview continues to be the most widely used method for selecting employees, despite the fact that it is a costly, inefficient, and usually invalid procedure.

A dozen years later, the recruiter Robert Martin said,

> Most of the corporate recruiters with whom I've had contact are decent, well-intentioned people. But I've yet to meet anyone, including myself, who knows what he (or she) is doing.

The new economy has taken notice. "In an interview you can tell if a person is a pleasant conversationalist, and you can give some technical questions to rule out the truly inept, but beyond that you might as well be rolling dice," wrote the founder of BitTorrent, Bram Cohen. Google's human resources head, Laszlo Bock, said it even more succinctly: "Interviews are a terrible predictor of performance."

What's so bad about an interview? The above critics were all too aware of some damning statistics. The evidence for the usefulness of job interviews is not unlike the evidence for extrasensory perception or alien abductions. There are some great anecdotes, but the closer you look at the data, the less compelling it is. In practice, job interviews appear to have little or no power to predict success on the job, beyond what might already be predicted from work experience or education.

Inevitably, interviewers favor candidates who "do well on interviews"—who look good, talk a good game, and make the right jokes. But doing well on the interview is not the same as doing well on the job. Of course, most interviewers insist they're aware of all that and correct for it—somehow. Most studies suggest that they don't correct for it enough. It may not even be possible to "correct" when much of the deciding is unconscious and automatic. People often get hired on hunches.

Behavior Predicts Behavior

The human resources profession has spent much of the past century trying to find better ways of evaluating candidates. One approach is by using *biodata*. A job applicant is asked questions about his past behavior, typically on a form, in the belief that the answers will help to predict how he will perform in the workplace.

Biodata is said to have originated in the insurance industry. At an industry convention in 1894, Colonel Thomas L. Peters of the Washington Life Insurance Company proposed asking a list of standardized questions for would-be insurance agents. Peters felt that by using the same actuarial analysis already in use for setting premiums, the company should be able to predict who would best be suited to the job. The premise of biodata is that "behavior predicts behavior." A person who has gotten five speeding tickets in the past year is likely to speed in the future and hence to be at higher risk for accidents.

A classic biodata question dates from World War II. The U.S. Naval Air Station, Pensacola, Florida, known as the Annapolis of the Air, was charged with training eleven hundred cadets a month, over ten times the peacetime figure. Not everyone had the right stuff to be a pilot. Training was grueling and expensive, and trainees were often ill for days—many had never been in a plane before. The war effort depended on accurately determining who had the talent and perseverance to succeed. Military psychologists designed a state-of-the-art questionnaire covering background, education, and interests. One psychologist at Pensacola, Edward Cureton, compared how recruits had answered the questions to how they subsequently fared in pilot training school. Cureton was astounded by what he found. One particular question on the list predicted success as a pilot better than the whole questionnaire did.

The question was, "Did you ever build a model airplane that flew?"

The recruits who answered yes were more likely to succeed as pilots. "That passion for airplanes, of people who had been doing it forever, ended up being predictive," explained Todd Carlisle, a psychologist in Google's People Operations division. "They would stick with it, no matter how many times they puked in the aircraft."

Biodata has gone in and out of fashion several times. Rightly or not, there's a perception that biodata is too crude a tool to provide much guidance in hiring "creative" innovators and managers.

That has limited its influence whenever and wherever employers have sought visionaries.

Creativity versus Intelligence

As a human resources concept, creativity is a legacy of the Sputnik-haunted Cold War epoch. The 1957 launch of the Soviet Union's globe-girdling satellite shocked Americans out of complacency. It was no longer a given that American enterprise led the world. Editorialists feared that the nation was falling behind in technical innovation. Universities revamped their curricula to emphasize science and creative thinking. Employers decided that they needed to jump on the creativity bandwagon, too. They began asking whether it was possible to spot the future inventors, entrepreneurs, and leaders.

The space race accelerated a trend that had already begun in psychology: the shattering of the concept of intelligence. For half a century, universities and employers had placed great faith in the notion of IQ. "Intelligence" was imagined to be a single quantity, responsible for all intellectual accomplishment and as measurable as blood pressure. Psychologists churned out IQ tests for a ready market of American universities and employers.

Were the sales pitches true, the task of employers should have been simple: hire the qualified applicants with the highest IQ scores. The reality was that IQ tests had little discernible value in hiring. That isn't to say that intelligence doesn't matter, but that work history and education convey much the same information and more. Equally troubling was the fact that high-IQ people aren't always good employees. Some are brainy slackers who never achieve much of anything.

The disconnect between intelligence and success began as an embarrassment to the IQ psychologists. By midcentury, they'd invented ways to turn this lemon into lemonade. One of Thomas

Edison's assistants, the Cornell University-trained engineer Louis Leon Thurstone, was so intrigued by the mysteries of intellect and success that he became a psychologist. Unlike other early proponents of IQ testing, Thurstone held that "intelligence" is not one thing but *many* distinct skills such as word fluency, spatial visualization, and reasoning. There wasn't much correlation between them, he argued. You can be brilliant at one thing, terrible at everything else.

J. P. Guilford, an army psychologist who segued to a career at the University of Southern California, took up Thurstone's thread with a vengeance. Guilford sliced and diced intelligence into as many as 180 distinct factors. In principle, all could be measured (if anyone had the patience, or cared).

A key figure in the "creativity" business was Ellis Paul Torrance (1915–2003), who decided that not only was creativity distinct from intelligence, but it was the really important thing. This was a savvy career move, coming just as disenchantment with IQ tests was peaking in the 1960s. Not only was there scant evidence that employment IQ testing did any earthly good, but the civil rights movement had made American businesses diversity-conscious for the first time. It was easily shown that IQ tests were biased against minorities, at least in the statistical sense. Employers dropped IQ tests, and other standardized personality tests, in droves.

We all know roughly what "intelligence" means. It's the ability to reason well and grasp the subtleties of the world around us. Intelligent people are quick thinkers who do well in university or business—provided they're motivated. "Creativity" is a more fluid term. Motivational speakers name-drop Leonardo da Vinci, Steve Jobs, Shakespeare, Henry Ford, Picasso, and Oprah Winfrey—all as examples of a fungible creativity. The business end of "creativity" tends to be equated with "success." But the story of many great successes challenges any simple identification.

The idea for Google came in a dream. Larry Page woke up one night with the concept, "What if we could download the whole Web, and just keep the links. . . . I grabbed a pen and started writing." What separates Page from all the PhDs who don't found a world-changing company? A lucky dream? Or something more?

Torrance attempted to address that sort of question. He began by immersing himself in a study of the lives of scientists, inventors, and explorers. A key issue he had to tackle is how creativity differs from intelligence. There were two main views on this, the disjoint and "nothing special" hypotheses. The disjoint hypothesis says that intelligence and creativity are completely different. You can have one, the other, both, or neither. No surprise there.

The "nothing special" hypothesis says that creativity is *nothing special*. Inside our heads, there's no deep distinction between creativity and intelligence. The distinction is strictly outside our heads. We look at certain consequences of thinking and ambition—Google, the lightbulb, cubism—and decide that there must be a special mental attribute that created them. But this is an illusion. It's plain old intelligence, motivation, hard work, and being in the right place at the right time.

By the "nothing special" theory, Page was perhaps lucky to have that dream. Had he not had it, or a similar brainstorm, he would not have cofounded Google. He doubtless would have been successful at something else, though maybe not in such a world-changing way. Thomas Edison was a font of "nothing special" sound bites. "Genius is one percent inspiration, ninety-nine percent perspiration," he said. Of course, someone less intelligent and driven than Edison or Page wouldn't have made much of those "lucky" flashes of inspiration. As another saying goes, you make your own luck.

It's easy to believe that there are elements of truth to both the "nothing special" and disjoint hypotheses. Torrance favored a middle position, the threshold hypothesis. This says that you have to be intelligent in order to be creative—but the opposite isn't the case. If you look at a random sample of creative, successful people, you will find that virtually 100 percent of them are highly intelligent. But if you look at a random sample of highly intelligent people, you'll find that few of them are creative or conspicuously successful in business or life. Put another way, Mensa meetings are filled with smart losers.

According to Torrance, the creative have an extra spark that distinguishes them from the great mass of the merely intelligent and educated. Torrance set out to find a way to identify that spark. By 1962 he had arrived at this précis:

> Creativity is production of something new or unusual as a result of the processes of:
>
> - sensing difficulties, problems, gaps in knowledge, missing elements, something askew;
> - making guesses and formulating hypotheses about these deficiencies;
> - evaluating and testing these guesses and hypotheses;
> - possibly revising and retesting them; and
> - finally, communicating the results.

This may strike you as both true and kind of obvious. Torrance was already monetizing his notion of creativity. He and colleagues devised the Minnesota Tests of Creative Thinking and the Torrance Tests of Creative Thinking. These standardized tests involved *divergent thinking,* a term coined by J. P. Guilford for what businesspeople have since come to call brainstorming. Guilford's classic test of divergent thinking was to *think of as many unusual uses for a brick as possible.* The more answers, and the more original the answers, the more creative the person was judged to be.

In similar tests with children, Torrance would supply a

stuffed bunny and challenge the kids to think of ways to improve the toy, to make it more fun to play with. For better or worse, these classic tasks reverberate through today's job interviews. Some companies actually ask the brick question. A more business-like update, used at Bank of America, is to have the candidate pull an unknown object out of a paper bag and devise an impromptu sales pitch for it. The "scoring," though informal, is much like Guilford's, with the number of distinct selling points tallied and extra credit given for originality.

Google interviewers test divergent thinking with this question:

? It is difficult to remember what you read, especially after many years. How would you address this?

This challenges the candidate to invent a new product on the spot. To answer such questions well, it's necessary to come up with many ideas, but also to edit and refine them. The complement to divergent thinking is *convergent thinking.* This is the process of using logic or instinct to narrow the range of possibilities—to decide which possible solutions best solve the problem.

It's easier to understand convergent thinking. We can express a logical proof in words. It's not so easy to articulate how "wild" ideas pop into one's head. Nor is it easy to make the ideas flow. ("I'm trying to think, but nothing's happening!") Divergent and convergent thinking are a yin-yang duality. Successful innovators need both. Those who excel only at divergent thinking may be flakes; those gifted with convergent thinking alone are intelligent but not creative.

In 1960 the University of Minnesota held a "Conference on the Space Age." The keynote speakers were the celebrity anthropologist

Margaret Mead and Ellis Paul Torrance. Mead remarked to Torrance that creativity had been studied before, with little result. "Why do you think it'll amount to anything this time?" she asked.

It remains a good question. Some psychologists still judge creativity research to be a slough with a few tufts of good footing (as Herman Melville said of philosophy). Torrance managed to create a new and hermetic specialty. The psychology of creativity has its own jargon and journals, with a set of canonical studies and lionized authorities. The hard question is whether today's understanding of creativity goes very far beyond common sense.

Today's psychologists usually define creativity as the ability to combine novelty and usefulness in a particular social context. The emphasis on social context is new, and it's especially relevant to "creativity" in the business realm. There is an individualistic, characteristically American way of looking at things that places much stock in the lone act of genius. The lone cowboy comes up with a million-dollar idea that inevitably wins over the world. But it doesn't always work that way. Peter Robertson invented a square-headed screwdriver that's better than the Phillips or slot kind. Just about every engineer agrees that Robertson's screwdriver is better, yet it never caught on except in Canada. No one's sure why.

Success is as difficult to explain as failure. Biz Stone, Evan Williams, and Jack Dorsey created Twitter, a fantastic success. It's just hard to say why, exactly. E-mail, text messages, blogs, podcasts, YouTube, Myspace, and Facebook all existed before Twitter. There were and are other microblogging sites. Twitter's added value is subtler, more contextual. It's a niche within an ever-shifting ecosystem of communication modalities. We apply the word *creative* to Twitter because it's successful. It's not so clear whether anyone, even the Twitter guys, could have anticipated that success. Innovators try interesting things and hope to catch a cresting wave.

Oxbridge and IBM

The psychological work provided a theoretical justification for something that was already going on: the use of puzzling and sometimes bizarre questions in personnel assessment. Applicants to Oxford and Cambridge have long been subjected to difficult admission interviews. "Oxbridge questions" include puzzles and philosophical paradoxes, often with a distinct air of whimsy. *Does a Girl Guide have a political agenda? How would you describe a human to a person from Mars? What percentage of the world's water is contained in a cow? Is it moral to hook up a psychopath (whose only pleasure is killing) to a reality-simulating machine so that he can "kill" as much as he likes?* And, to Cambridge theology students, this poser: *Could there be a Second Coming if mankind disappeared from the planet?*

In the United States, the computer industry was particularly receptive to brainteaser interview questions. That's often traced to IBM. One of its legendary engineers, John W. Backus, was a human resources department nightmare, a man whose many talents defied measurement. After dropping out of the University of Virginia, Backus was drafted into the U.S. Army during World War II. The army gave him a battery of aptitude tests and concluded that he was too brilliant for regular service. Instead, he was sent back to university at public expense.

Backus went on to pursue a master's degree in mathematics at Columbia University. By pure serendipity one day he strolled past IBM's Madison Avenue headquarters. It was displaying one of the firm's new electronic calculators, a marvel of miniaturization about the size of a Manhattan office. As Backus was gazing at the thing in wonderment, an IBM tour guide began asking him questions. Backus mentioned he was studying maths. The guide beckoned him upstairs for what turned out to be a job interview. It consisted of a series of logic puzzles.

The year was 1950, and IBM was in a quandary. It was coming to the belated realization that software design was not electrical engineering at all, but a whole new field, lacking a name or a suitable degree program. Even the term *software* did not exist, and *hardware* meant wrenches and toilet snakes. The people IBM needed to hire might conceivably come from any background. Asking logic puzzles was an attempt, however makeshift, to identify those capable of thinking in new ways.

Backus did well enough on the puzzles to be hired on the spot. He eventually led the team that produced FORTRAN, the first high-level computer language. (Its importance to software has been compared to the transistor's importance to hardware.) Since no one had experience or education relevant to high-level programming languages—because they didn't exist—Backus had to cast his net widely. "They took anyone who seemed to have an aptitude for problem-solving skills—bridge players, chess players, even women," said Lois Haibt, a newly-qualified maths graduate. The team grew to ten and included a crystallographer and a code breaker. Backus described his creative process in terms much like Edison's or Torrance's: "You have to generate many ideas and then you have to work very hard only to discover that they don't work. And you keep doing that over and over until you find one that does work."

In 1957, William Shockley, the most cantankerous of the three men credited with inventing the transistor, moved west to build and market electronics. His Shockley Semiconductor Laboratory, the first Silicon Valley start-up, was in Mountain View, a bike ride from where the Googleplex now stands. Shockley was so nuts about using logic puzzles in hiring interviews that he timed applicants with a stopwatch. Maybe that should have been a tip-off. Shockley was a holy terror to work for. Mere months after they were hired, eight of his brightest employees—the "Traitorous Eight"—got so fed up they resigned. They went on to found companies like Fairchild Instruments and Intel. Ever since, brainteaser interviews have been part of hiring in the computer industry.

Selling Sergey's Soul to the Devil

In July 2004, two enigmatic billboards went up on opposite sides of the country. One was in Harvard Square; the other, off Highway 101 in Silicon Valley. Each billboard was stark black text on a white background, reading:

{first 10-digit prime found in consecutive digits of *e*}.com

There was no mention of who had put up the billboards or what was being advertised.

It was a test. As expected, the billboards drew publicity. A gaggle of mathematically inclined bloggers wrote about the billboards; then a radio programme did a piece on the mystery. One of the first to solve the puzzle was the iconoclastic physicist and mathematician Stephen Wolfram. Born in London in 1959, Wolfram had been a child prodigy and had published an important paper on quarks at the age of seventeen. Three years later he had a PhD in particle physics. In the 1980s, Wolfram received a MacArthur grant, worked at the Institute for Advanced Study, and collaborated with Richard Feynman. In 1987 he cofounded Wolfram Research to market Mathematica, the globally used calculating program for scientists and engineers. It took Wolfram just a single line of Mathematica code to solve the billboard's puzzle.

Let me explain what the billboard was asking. Start inside the brackets with the italicized lowercase letter *e.* This is Euler's number, approximately 2.71828.... One way to explain *e* is to say that it's a measure of the power of compound interest. Borrow a quid from a loan shark charging 100 percent interest, compounded daily, and you will owe just under *e* pounds at the end of the year—£2.72.

Compound interest is only one of *e*'s guises. It is an almost mystic number that turns up in diverse mathematical contexts (most having nothing to do with usury). In that respect, *e* is something like the better-known number pi, which appears in formulas having nothing to do with circles. Like pi, *e* cannot be expressed exactly in decimal notation. It's an endless string of digits, starting with 2.71828... and never repeating.

That's how it fits into this puzzle. Because the digits of *e* never repeat, you can expect to find any sequence of digits in *e* if you look long enough. Your phone number is in *e,* somewhere. So is everyone else's phone number, credit rating, and batting average. The world's population, tomorrow's winning lottery numbers, and the current temperature in Tangiers are all in there.

The billboards were asking for the first 10-digit *prime* number in the digits of *e.* A prime number is any that cannot be divided by any other number greater than 1. Seven is a prime number, and so is 23. Eight is *not* a prime number because it can be divided by 2 and 4. Nor is 25, which divides by 5. It's been known since Greek antiquity that the distribution of primes fits no neat pattern. There are prime numbers of all sizes.

Many ways are known for determining the digits of *e,* and for identifying prime numbers. To us in the twenty-first century, the easiest method by far is to Google it. Hundreds of sites will give you *e* to more digits than you know what to do with. Other sites will give you lists of prime numbers.

This isn't as helpful as you might hope. If you wanted to

locate the first occurrence of a specific *short* number (like your weight) in *e*, all you'd have to do is to navigate to a page listing the digits and use the browser's search function to find the desired number. Unfortunately, there are *lots* of ten-digit prime numbers: more than four hundred million of them. You would have to test each one, like a thief trying every combination on a lock. Even with a helper supplying prime numbers, and supposing you could test a new one every second without sleep, it would take nearly fourteen years to run through all the possibilities.

The only realistic way to solve the puzzle is to write code. That's what Wolfram did. Fortunately, Mathematica is streamlined to perform the sort of number theoretic calculations needed. Wolfram's single line of code was

```
Select[FromDigits/@Partition[First[RealDigits[E,10,1000]],
10,1],PrimeQ,1]
```

This promptly identified the ten-digit prime number as 7,427,466,391. It starts ninety-nine spaces to the right of *e*'s decimal point:

> 2.71828182845904523536028747135266249775724709369995
> 95749669676277240766303535475945713821785251664274
> 27466391...

The billboard placed a *.com* after the brackets, so Wolfram typed "7427466391.com" into a browser. It took him to a page that said

> Congratulations. You've made it to level 2.

It was followed by a second puzzle. Answering it correctly qualified the solver for the third level. Soon, very smart people all over the world were tackling the puzzles. As with a video game, the number of players dwindled with each succeeding level. Finally, at the end of a succession of puzzles came a prize (or a MacGuffin): an invitation to send a CV to Google.

Perhaps no company has done so much to popularize the use of intellectual challenges in hiring as Google. As a Russian-immigrant child in Maryland, Sergey Brin spoke little English, delighting instead in mathematical puzzles. Young Larry Page was fascinated by the eccentric inventor Nikola Tesla. At the University of Michigan, Page built a working ink-jet printer out of LEGO blocks.

In Google's first five years, Page or Brin or both interviewed each candidate. Even today, Page still signs off on every hire. The founders were known for novel interview questions. When Alissa Lee, an attorney, was interviewed, Brin asked her to draw up a contract to sell his soul to the devil. It had to be e-mailed to him in the next thirty minutes.

"Amid the surreal oddity of it all," Lee said, "I had forgotten to ask him all sorts of lawyerly questions, like what sort of protections he needed, what conditions he wanted to attach, and what he wanted in return for his soul. But then I realized that I had missed the point. He was looking for someone who could embrace a curveball, even relish it, and thrive in the process of tackling something unexpected." Lee was hired.

Google's billboards were not quite what they appeared. The company was already deluged with CVs. The smart people who did solve the puzzles, and wanted a job, did not get any preferential treatment. The billboard stunt was primarily a way to burnish Google's image as an innovative workplace. For a pittance, the ads earned Google old-media attention and viral marketing on the web. Not everyone paid attention, and many who did must have concluded that Google was brimming with hopeless geeks. Yet almost everyone with the qualifications to work at Google *did* hear about the billboards, one way or another. They commanded the attention of world-class geniuses who didn't need jobs, like Wolfram. Mainly, the billboards helped ensure that when software engineers looked for their next job, they thought of Google first.

The Church of Apple

If there's a company that's even cooler than Google, it's Apple. The problem with getting hired at Apple is "the amount of people that would cut off their own testicle for a job," one applicant reported. He was applying for a position at an Apple store opening in Florida—as a salesperson earning about eleven dollars an hour. He didn't count on the Church of Apple:

> Have you ever gone to a Church, where everyone is totally convinced in the idea of a loving God watching over us all? If so, you can probably relate to the interview process of Apple. The entire interview process took place over a two- to three-month period for a new store opening. It consisted of an introduction to the company, where four or five vested employees preach the goodness of Apple products and how life changing they are, then you're asked to stand up and introduce yourself, and then, after that, it's time to dance like a performing monkey to the hiring manager's approval.... The entire process really smells like you're being interviewed for a pyramid scheme.

You've probably noticed that the Apple store staff is as carefully cast as the help at Disneyland. Everyone fits the role perfectly. No one is uncool. That's because they turn an awful lot of people away. When Apple opened its store on Manhattan's Upper West Side in 2009, it got ten thousand applications and hired just over two hundred of them (about 2 percent). One of the questions posed in Apple's group interviews speaks volumes about the corporate culture: "What happened in 2001?" Mention 9/11, and you'll be coolly informed that there are other good answers. The "correct" responses: "The iPod was introduced!" and "The first Apple store opened!"

This is one way of hiring for that subset of companies that can pull it off. The coolness ensures that the company will get dozens of applicants for every opening and that those applicants will submit to a gauntlet of interview puzzles, stunts, tests, and hazings. At the end, the interviewers skim the most dazzling talent off the top and reject everyone else (most of whom may be perfectly well qualified). It's not hard to understand why companies do this. The mystery is why so many job seekers are mesmerized by corporate mystiques.

It wasn't so long ago that job hunters scanned the want ads of their local newspaper, getting ink on their hands. Most of the employers advertising were local. The Internet and Monster.com have opened job seekers' eyes to a world of options beyond their own cities. Features comparing company benefits and cultural intangibles draw floods of applications to the highest-rated companies. For these companies, buzz has the same effect it does on nightclubs. The popular ones are very hard to get into.

In Silicon Valley, outlandish benefits have about as long a history as tricky interview questions. Hewlett-Packard was one of the pioneers, offering free snacks, and expensive gifts for newlyweds and new parents. Many of Google's perks were cribbed from other companies like Genentech (the informal Friday meetings) and Facebook (bring your dog to work). Today, free chef-prepared meals are the norm at Silicon Valley firms, and having a child is worth a mid-four-figure sum. As Larry Page said, "Our competitors have to be competitive on some of these things."

If this is generosity, it's the kind that Ayn Rand could have loved. When Google's early investors were leery of the free food, Sergey Brin supplied a characteristically quantitative defense: Employees would otherwise have to drive to lunch, wait to be served, and drive back. Eating on campus saves thirty minutes per employee per day. Based on that, the food pays for itself.

More to the point, there's evidence that the companies that people most want to work for are "better" by almost any measure. Alex Edmans computed that a portfolio of *Fortune* magazine's "100 Best Companies to Work For" from 1984 to 2005 outperformed the market by 4.1 percentage points a year.

Why? Robert Levering and Milton Moskowitz, the authors of *Fortune*'s annual lists, have long argued that the crucial ingredient to their best workplaces is trust. It's no secret that we live in a cynical age. How much time at a typical company is spent joking about the boss? More work gets done at those rare businesses where people care about the product, their superiors, and the company itself.

Page recalls a grandfather, a worker at Chevrolet's Flint, Michigan, plant, who carried a leather-clad iron pipe to work. The pipe was to protect him from company goons during strikes. Page, who aspired to start a company from age twelve, early concluded that unhappy workers are unproductive workers.

Companies like Google and Apple go to great lengths to present themselves as creative, enlightened workplaces. The novel benefits may be the least of that, but they are one high-profile way of showing that management values its human capital. In Silicon Valley, at least, benefits weren't cut much when the economy tanked, and perks seem to be making a comeback. Some of the smaller gaming and social network firms now make Google look stuffy and a little cheap. New employees at Asana, a software firm, are treated to a $10,000 spending spree on computer and electronic gear. It sometimes seems that firms are hard put to find something new to dangle before prospective employees. After work hours, Scribd converts its San Francisco offices into a go-cart track. There's also a zip line. Zynga, the social network game company, promises to send someone out to employees' homes to wait for the phone or cable guy. "We do have a workforce for whom this is their first job," noted Zynga chief people officer Colleen McCreary. "I worry if they ever wanted to go work somewhere else."

Perhaps it's not surprising that employers complain of a sense of entitlement among many job seekers—jobless recovery or not. A recent study found that 41 percent of new 2010 graduates turned down job offers, the same percentage as in the boom year of 2007. And when interviewers ask whether the candidate has any questions, perks are often a topic of intense curiosity. One interviewer tells of a job hunter who showed unseemly interest in the company's legal benefits. "He wanted to know if it covered the cost of filing lawsuits, if it covered him if he got sued himself, if it applied to any lawsuits he currently was involved in, and if he could 'theoretically' use it to sue the company itself."

Some of today's interview questions are as much about screening out the big spoiled babies as finding geniuses. Rakesh Agrawal, a consultant who's worked for Microsoft, AOL Search, and the *Washington Post*'s website, likes to ask job candidates what they think of his company's product. Once, Agrawal ran into an applicant socially and helpfully said it would be a good idea to check out the product before the interview. On the appointed day, the candidate confessed he hadn't tried the product. He explained that he had looked at the website and decided the product didn't interest him.

Said Agrawal, "He was interviewing to be VP of marketing."

QUESTIONS

Traditional Brainteasers

Here are some traditional brainteasers long used in technology company interviews. Today, you'll find them at many other companies, too. (Answers start on p. 144.)

- **?** There are three men and three lions on one side of a river. You need to carry them all to the other side, using a single boat that can carry only two entities (human or lion) at a time. You can't let the lions outnumber the men on either bank of the river because then they'd eat them. How would you get them across?
- **?** Using only a 4-minute hourglass and a 7-minute hourglass, measure exactly 9 minutes.
- **?** Find the minimum number of coins to give any amount of change. (From a U.S. company, so think: 1¢, 5¢, 10¢, 25¢, 50¢.)
- **?** In a dark room, you're handed a deck of cards with *N* of the cards faceup and the rest facedown. You can't see the cards. How would you split the cards into two piles, with the same number of faceup cards in each pile?
- **?** You're given a cube of cheese and a knife. How many straight cuts of the knife do you need to divide the cheese into twenty-seven little cubes?
- **?** There are three boxes, and one contains a valuable prize; the other two are empty. You're given your choice of a box, but you aren't told whether it contains the prize. Instead, one of the boxes you didn't pick is opened and is shown to be empty. You're allowed to keep the box you originally picked ("stay") or swap it for the other unopened box ("switch"). Which would you rather do, stay or switch?
- **?** You're in a car with a helium balloon tied to the floor. The windows are closed. When you step on the accelerator, what happens to the balloon—does it move forward, move backward, or stay put?

Three

Punked and Outweirded

How the Great Recession Mainstreamed Bizarre Interview Questions

It's said there was a time when U.S. company Walmart had to hire anyone with a pulse. That joke has long passed its expiration date. In this grim economy, Walmart is flooded with applications for every opening, many from the painfully overqualified, leading Walmart's interview questions to become more thought provoking. One example is this one: "What would you do if a customer came into the store not wearing the appropriate amount of clothing?" This may not be as difficult as Google's questions are known for, but it's a subtle psychological puzzle resisting a glib comeback. A good answer would allow that there is more than one kind of possible fashion offender. Teenagers wearing provocative new fashions considered cool in their peer group are distinct from mentally disturbed adults in ill-chosen dress. Walmart likes to see candidates who aren't limited to a "one size fits all" response.

The retail giant has reason to be choosy. In September 2009, the U.S. Labor Department reported that job seekers outnumbered openings six to one. These unemployment numbers have spread riddles, loaded questions, and multiple-interview marathons across the corporate food chain, into mature and less cutting-edge industries. The same is true in the UK.

"If you could be any superhero, who would it be?"
"What color best represents your personality?"
"What animal are you?"

These questions aren't from some wacky Silicon Valley start-up—they're asked at AT&T, Johnson and Johnson, and Bank of America, respectively. Today's employers feel a not unreasonable obligation to make the most of their unprecedented ability to be selective. Unfortunately, there's still no foolproof way to identify the most talented and motivated employees. In mainstream companies as well as the tech field, unusual and occasionally deceptive techniques are turning up in the name of gauging "creativity" or "culture fit."

"Most people don't interview very frequently," explained consultant Rakesh Agrawal. "They do it maybe twice a year. You do the kinds of things you've heard about, and it perpetuates." Weird interview questions are a meme, like a joke or viral video. It's catchiness, rather than proof of their effectiveness, that keeps them in circulation.

Screeners and Litmus Tests

As the job market collapsed in 2008, employers took to career-fair or phone interviews in which the interviewer poses so-called screener questions or litmus tests. These are simple questions or criteria that (supposedly) weed out the "wrong" people. That's said to be necessary with today's profusion of applicants. Screeners may test work-related knowledge, motivation, personality, fit with the corporate culture, and ability to handle stress. Often it's hard for the applicant to tell what the point is. But those who give "wrong" answers generally don't make it to the next stage of interviewing.

Rakesh Agrawal asks applicants to name their favorite Internet product. His follow-up is, "How would you improve it?"

The follow-up is a bullshit detector. "I've had people tell me they were really passionate about Gmail, and then tell me about features that they wanted, that had been in the product from day one."

Many companies ask trivia questions about the firm itself. It's thought that a truly motivated applicant will have done some research. Goldman Sachs interviewers ask candidates the firm's stock price. Johnson and Johnson sometimes asks for the largest lawsuit currently pending against J&J—a wry acknowledgment of the realities of the pharmaceutical business.

Morgan Stanley asks interviewees to name a recent story they read in the *Financial Times*—apparently, a lot can't—or to give the square root of 0.01 (it's 0.1). JP Morgan Chase asks the value of pi. It's believed to be instructive to see how many digits the candidate can recite.

Bloomberg LP is big on proofreading. Some applicants are given a test in which they have to count the number of times a given letter appears, uppercase or lowercase, in a paragraph. It's a lot harder than it looks. (Don't believe it? Count the *h*'s in this paragraph. There are fifteen and hardly anyone gets all of them.)

On Wall Street particularly, interviews can verge on hazing. At Bloomberg, interviewers may interrupt candidates as they try to answer. "Everything I said they responded 'Are you sure? Are you sure?'" one Bloomberg candidate recalled. "When writing the code, they would question it and laugh, trying to trip you up, even when you were right."

Peter Muller, the manager of Morgan Stanley's hedge fund PDT, is famous for asking job applicants to estimate the amount of cash in his wallet "with 95 percent confidence." The candidate is supposed to name two figures, a low one and a high one, and be 95 percent sure that the actual amount is within that range.

Being cautious, people typically pick zero as the low figure (a guy like Muller might be beyond carrying cash) and maybe $500 as the high. Immediately, Muller pulls $500 out of his wallet.

Would you care to revise that guess? They do, and whatever high figure they say this time, Muller finds it in his wallet—like a magician pulling quarters out of the air.

"If you were a cartoon character, which one would you be and why?" This is a question that Bank of America has asked aspiring personal bankers. "I said Yogi Bear," one applicant recalled. "I can't remember the reason why I said this, but the hiring managers were all applauding it." He was given the position immediately.

Questions like this, sometimes downright silly, have become more common. Since being a maths or tech whiz is irrelevant to running most businesses, mainstream companies have redoubled their efforts to find the perfect match of candidate and corporate personality. Job interviews have become like speed dating. Whole Foods interviewers have candidates describe their perfect "last meal." It's a quick way of gauging the applicant's knowledge of and passion for food. Expedia does the same thing with travel, asking questions like "If you could go camping anywhere, where would you put your tent?"

Schlumberger, the oil field services company, has a very clear idea of the personalities it's seeking: extroverted engineers. Perhaps they think that anyone sent to a succession of remote outposts will go nuts unless he can make friends quickly. Schlumberger's interviewing process is therefore structured around weeding out the wallflowers. Hence overnight recruiting trips where candidates share beers with managers and are asked questions like "What are your hobbies?" "Reading the latest Jonathan Franzen novel" is not a particularly good answer.

There is much speculation about how best to answer the "scale of 1 to 10" questions that have become popular lately. For example, Wells Fargo has candidates rate their competitiveness on a 1-to-10 scale. In practice, almost everyone rates himself 8 or higher, and if you want the job, you should avoid false modesty.

Online retailer Zappos has a trickier question: "On a scale of 1 to 10, how weird are you?" The preferred answer there is somewhere in the middle, CEO Tony Hsieh explained. A 1 is "probably a little bit too straight-laced for us," and a 10 "might be too psychotic."

It's possible to fail a litmus test without realizing it. According to one recruiter, Nordstrom screens out more than 90 percent of female applicants with a simple three-part test:

- Is the applicant wearing black?
- Is she wearing heels?
- Is she wearing a watch?

All three answers should be yes. It's as arbitrary as all get-outs. But given that just about everyone in American retail is dying to work for Nordstrom (ranked #53 on *Fortune*'s 2010 list of "100 Best Companies to Work For"), some hirers feel there's little reason to consider those who don't get their dress code right.

Equally covert is the most ubiquitous personality measure, the "airport test." After meeting with the applicant, the interviewers have a postmortem on his or her general likability. As Larry Page explained it, "Just think about if you got stuck in an airport with this [job candidate], on a long layover on a business trip. Would you be happy or sad about it?" They want to hire people who are fun to be around.

Today the two most widespread litmus tests of all must be credit ratings and employment status. Many employers don't want to hire those with bad credit or those who are unemployed. Why? The theory is that bad credit reveals poor judgment—not just in the mall but in the workplace. "If you see a history of bad decision-making, you don't want that decision-making overflowing into your organization," said Anita Orozco, a human resources director with the chemical company Sonneborn.

Even more Machiavellian is the practice of not hiring the unemployed. This is based on the presumption that companies naturally retain the "best" people when they lay employees off. Therefore, the top workers are concentrated in companies that have had layoffs, and you want to hire them—not anyone who actually needs a job. "Most executive recruiters won't look at a candidate unless they have a job, even if they don't like to admit to it," Lisa Chenofsky Singer, a human resources consultant specializing in media and publishing jobs, told CNNMoney.

Those lucky enough to get face time with an interviewer can expect...more interviews. "We're definitely putting people through more paces than ever before," said Michelle Robinovitz, a recruiter for Aarons Grant & Habif, an Atlanta accounting firm that regularly heads local lists of best places to work. "In better times, we did one or two interviews. Now we really want to make sure someone will fit and we do a minimum of four interviews." Robinovitz predicts this trend will survive the recession. Companies have learned that they need to be "streamlined" and that bad hires are costly.

Not every company can afford to fly in candidates for a day of intensive interviewing, as Google does. More often job seekers face a worse torture, the Kafka interview. They are called back repeatedly, for an ever-indeterminate series of interviews that may end in a job, a rejection—or not even that. Sometimes the callbacks simply cease without so much as an e-mail of thanks. Candidates may face half a dozen interviews on half a dozen days and still have no clue where they stand. By the old standards, five callbacks would have indicated interest. Today it can mean nothing.

The suspense doesn't always end with the interviews. It's increasingly popular to offer promising candidates a several-month trial position (with few or no benefits). At Google, this is known as being a contractor. Guess what? The "job" is really another interview. Only after the term is up does the company decide whether to offer a permanent job. This wouldn't fly with an applicant who already had a decent job, of course. But today's zombie hordes of unemployed and underemployed are willing to claw at anything that even looks like a job.

Do Unconventional Interview Questions Work?

The unanswered question is whether today's ever-more-polymorphous interviews succeed in identifying better employees. The use of peculiar questions and arbitrary tests may seem to go against one of the few rock-solid precepts of today's human resources profession. This holds that any method of selecting job candidates should be as closely related to the work as possible. Most HR people place the most faith in work sampling, where the candidate is asked to perform or simulate work similar to that which he'd be doing if hired. Selling Sergey's soul to the devil was an example, if an offbeat one. Statistical studies of work sampling (a famous one was done by AT&T from 1956 to 1965) showed impressive predictive ability.

The usual justification for "creative thinking" riddles and personality assessments is that they test broad, general abilities, not tied to a specific set of skills. Whether they do that is hard to say. What's certain is that "pet" questions take on a talismanic quality for some interviewers. Just as athletes don't change their shirt during a winning streak, interviewers keep asking the same questions because of a few remembered instances where it supposedly "worked." The fact that many of the most admired, innovative companies use such interview questions seems to speak for itself ("You can't argue with success").

It's far from clear that either reason holds water. The human resources profession is full of customary practices of no demonstrable value. The psychologist Daniel Kahneman tells the tale of a test once used by the Israeli military to identify candidates for officer training. A group of eight recruits, stripped of insignia, was instructed to carry a telephone pole over a wall without letting it touch the wall or the ground. The point was to observe who took charge (the "natural leaders") and who fell meekly into place behind them (the "followers"). "But the trouble was that, in fact, we could not tell," Kahneman said. "Every month or so we had a 'statistics day,' during which we would get feedback from the officer-training school, indicating the accuracy of our ratings of candidates' potential. The story was always the same: our ability to predict performance at the school was negligible. But the next day, there would be another batch of candidates to be taken to the obstacle field, where we would face them with the wall and see their true natures revealed."

Similar tactics are alive and well throughout corporate America. In today's overheated job market, a common test is to seat a group of candidates for the same job around a conference table for a "group discussion." They know that only one will get the job. The discussion becomes a little reality show, with the recruiter quietly noting who takes charge. It's doubtful that it works any better than the Israeli army test did.

Proving that a hiring technique works—or that it doesn't work—is a complex exercise in statistics. Were one to demand that a hiring criterion be 100 percent reliable, employers would have to hand out jobs at random. There aren't any 100 percent reliable criteria—not work history, not grades, not anything. Hiring is always a game of chance. Many job seekers complain that some talented people do poorly on today's unconventional interview questions—ergo no one should use them in deciding whom to hire. This isn't a compelling argument for the reason given above. But psychological studies indicate that people are apt to view almost *any* criterion as "unfair" when it's used to decide who's hired or promoted. The sense of unfairness is greater when the criterion is unfamiliar. A traditional job interview is a conversation. The job offer or rejection comes days or weeks later, affording a certain emotional distance. Creative-thinking questions often bring the rejection right into the interview, right in your face. If you fail, you generally know you've failed. That feels worse than a rejection days later. Admittedly, this attitude may not make sense, but when have emotions ever had to make sense?

QUESTIONS

The Numeracy Screen

Businesses that deal with numbers often pose short, tricky maths questions in their initial phone interviews. You can be sitting at a computer as you answer, but it doesn't always help. Often a pencil and paper is more useful. (Answers begin on p. 162.)

? According to a survey, 70 percent of the public likes coffee, and 80 percent likes tea. What are the upper and lower bounds of people who like both coffee and tea?

? At 3:15, what is the angle between the minute and hour hands on an analog clock?

? How many integers between 1 and 1,000 contain a 3?

? A book has *N* pages, numbered the usual way, from 1 to *N*. The total number of digits in the page numbers is 1,095. How many pages does the book have?

? How many 0s are there at the end of 100 factorial? [That's 100 multiplied by every whole number smaller than itself, down to 1.]

Four

Google's Hiring Machine

How They Pick the One to Hire out of the 130 Who Apply

Everyone knows Google's doing a good job at hiring smart people," Amazon hiring manager Steve Yegge wrote in a much-read 2004 blog post.

> It's not just anecdotal; the numbers speak for themselves. We lose a lot of our best candidates to them. . . . What's not so clear, I think, is that Google is actually so good at technical recruiting that it's not just a difference in magnitude; it's a difference in kind. What they're doing can hardly be called recruiting anymore. The term "recruiting" implies that you're going out and looking for people, and trying to convince them to come work for you. Google has managed to turn the process around. Smart people now make the pilgrimage to Google, and Google spends the bulk of their time turning great people away.

What are all those great people looking for? It's not the money, even though Google churns out employee-millionaires. (The first thirty Google employees received stock worth half a billion dollars by 2008. That's half a billion for each of the thirty.)

What Google offers is more like an elite university or think tank. But universities are about theory, and Google is about practice. It offers the heady challenge of creating the new digital universe. In Yegge's analysis, "Smart people go where smart people are, which enables them to launch cool stuff, which attracts more attention, and suddenly you have a feedback loop."

A Signal in a Lot of Noise

At Google, human resources is known as People Operations, or People Ops. Todd Carlisle, a youthful industrial psychologist with shoulder-length hair, began working for Google's People Ops in 2004. "They had a lot of data," he explained, "and they didn't have anyone to look at it, analyze it, and tell them what it meant."

It was Carlisle's job to perform statistical analyses to determine what factors matter in hiring. "The founders are engineers, and they're used to looking for a signal in a lot of noise," he said. Yet when it comes to human beings, the statistical approach runs up against resistance. "It's like a computer telling you this is the person you should marry. Everybody feels like they know, when they do an interview, what they're looking for. I sat people down and asked them, 'What are you looking for?' and I got totally different responses from just about everyone. So I thought, 'All these people can't be right.' "

Carlisle has explored biodata techniques at Google. "I started to delve into things like, 'When did you get your first computer?' " he said. The earlier the applicant had been exposed to computers, Carlisle found, the better the applicant performed at Google, as measured by quarterly performance reviews and other criteria. Another predictor was an update of Cureton's model airplane question: "Did you ever make a computer from a kit?" Like building model airplanes, making computers from kits may seem geeky. But the people who did it generally have a lifelong passion for computers. That passion augurs well for surviving in an intense environment where everyone is obsessed with all things digital.

In 2006, Carlisle devised the Google Candidate Survey. It was a homegrown personality test designed to measure culture fit ("Googliness") of potential staff. The company began by asking every employee who had been at the company at least five months to fill out a three-hundred-question survey. The results were compared to statistical measures of performance at Google. As expected, Carlisle found that many things had no effect on performance, but a few did. The test was gradually refined and polished, and by 2007, Google was asking job candidates to fill out the survey.

For all the emphasis on brains and ambition, Googlers believe in an open, collaborative environment that goes against the stereotype of engineers as driven loners. The Googleplex is a gregarious place. The cubicles have low walls so that no one is truly isolated—or as an outsider might observe, no one has any privacy. Larry and Sergey shared an office for much of the company's history. Ironically, Googlers craving some temporary quiet time seek the refuge of an empty conference room.

It is therefore vital that staff be able to thrive in what some might think of as a fishbowl environment. "We like people to be very collaborative and to understand that everything they're building, they're building as a team," Carlisle explained. "You can't just work on your piece of code and assume it's fine; it's got to work with other people's code."

How do you spot extroverted software engineers? The simplest way is to ask people how much they like collaborating. That runs into the oldest problem of biodata: people say whatever they think the employer wants to hear.

A solution is to design questions so that it doesn't matter so much whether people misrepresent. One question on the Candidate Survey ran,

> Please indicate your working style preference on a scale of 1 to 5.
> 1 = Work Alone: A personal espresso machine and a box of Toblerone on my desk and I'm raring to go!
> 5 = Work in a Team: Ten voices at once, egos colliding... ah, the challenge of getting a word in edgeways!

This question's wording is weighted against the answer Google wants. It reminds the applicant that teamwork can be unproductive, while linking solo work to upscale coffee and chocolate. The use of the 1-to-5 scale promotes candor. Not many people give the extreme answers of 1 or 5. They realize an answer in the middle is "safer." Yet loners tend to answer 2 or 3, while gregarious types answer 3 or 4. There is a statistical difference between the personality types, even though many are shading the truth.

"One of the things I tested for was, if you've done coding competitions—and I tested for specific ones—how well do you do once you get here?" Carlisle said. Google sponsors one of the best-known competitions, the Google Code Jam. Not a few engineers dreaming of a job at Google enter. "What I actually found was that people who had participated in them, once they get here, are even less successful than people who've never been in them," Carlisle reported.

Why are the world's top competitive coders not especially good fits at Google? The statistics don't answer that. Apparently, it's a little like a fashion-modeling agency looking for beautiful women. It doesn't necessarily follow that winners of beauty contests would make the best models. Carlisle theorizes that the nature of coding competitions—one individual against the world, working on a narrowly defined project with a beginning and an end—has little to do with Google's collaborative environment. The people who enter competitions may want to vanquish rivals in a short time frame; at Google, they'd probably be bored.

The Google Candidate Survey has mostly been phased out. It was found that almost nothing qualified as a surefire, all-purpose predictor for success at Google. The items that worked for ad sales did not work for software engineers or public relations. But particular departments sometimes ask questions from it in interviews. The takeaway: expect some questions about work style and personality. Google is still very keen on smart people who are natural collaborators.

"The Package"

In most companies, information about a candidate comes in in dribs and drabs. Employers overvalue the first few pieces of data they encounter. Thereafter they tend to ignore information that doesn't fit their initial impression. This is a classic fallacy of decision making. In order to avoid it, Google believes in collecting all the information about the candidate before presenting it to decision makers. For this reason, its hiring is strongly centralized. Information on those being considered for the Mumbai or Wrocław offices is funneled back to Mountain View. Overseas candidates can expect a video interview with someone at the Googleplex.

The supreme embodiment of this philosophy is "the package." It's a forty- or fifty-page dossier on each Google applicant, explained Prasad Setty, Google's director of people analytics and compensation. The package is a biography containing all the information that Google has been able to gather about the applicant. The company is effective at "Googling" people, figuratively and literally. In general the package has academic grades; the candidate's CV; work samples (anything from published papers to press releases to shipped products); reports of references; and information from the web, which could include blog posts or even social network postings.

The package has spawned a variety of urban legends about Google's hiring standards. It's claimed that you can't get hired at Google unless

- you had a GPA of 3.7 or higher (3.0 for nontechnical positions);
- you attended Stanford, Caltech, MIT, or an Ivy League university;
- you got triple 800s on your SATs; and/or
- you have a PhD.

None of these is a must. But you'll be competing against applicants who have many or all of these qualifications.

"Google was my first job out of university," recalled one now-former employee. "I was an English student at a prestigious university and was hired to work in HR. That is one of the problems I had with Google right there—is it really necessary to hire Ivy League graduates to process paperwork? I went from reading Derrida to processing Status Change Request Forms for employees to go on paid leave."

Many find Google's concern with universities and grades exasperating. The *New Yorker* journalist Ken Auletta termed it "preposterous." Roni Zeigler, a physician with an advanced degree in medical informatics, recalled his surprise at having to supply secondary school grades during the vetting. (He was hired.) Tech bloggers claim that Google ignores CVs not from top universities.

Google's hirers insist they have been misunderstood. Since Google asks for grades, and some other companies don't, outsiders conclude that Google has an overweening or naive faith in grades. In fact, the goal is to give grades neither more nor less weight than they deserve. "Last week we hired six people who had below a 3.0 G.P.A.," the head of People Ops, Laszlo Bock, bragged in 2007. Carlisle (who got his PhD at the non-Ivy Texas A&M)

said Google uses a background from a top university as a signal, given "that someone has done that vetting for us, but we certainly don't use that to exclude anybody. We often look at people who have overcome some adversity to get to where they are," he said. "Are you the first one in your family to have gone to university? I hired somebody the other day who was not only the first in her family to go to university but worked full-time in university so that her sister could go as well. We hired that person, and she went to a very non–Ivy League school."

It might be more accurate to say that Google acts like an Ivy League university in evaluating candidates. Its policy could be described as affirmative action for the very, very smart. "We do go out of our way to recruit people who are a little different," Larry Page once said. Carlisle therefore sees his role as "finding the people we would ignore. Like, who's the kid in his village in India who repairs all the phone lines even though he's twelve, because no one else can do it, and he's got that technical aptitude?" he elaborated. "Who are these girls in the inner city of Detroit who have this technical aptitude, and how can we find them when they're younger and make sure they come here?"

The proportion of female employees at Google is now said to be close to 50 percent. This is impressive in a society that's still skewed against women scientists and engineers. Google hires from all over the globe, and many come to Mountain View, making the Googleplex a cosmopolitan place.

The Rule of Five

As at other tech companies, Google's hiring interviews are conducted by peers. People Ops advises the interviewers on everything from fair employment law to the "art of rejection." "They actually go through a training course on how to let candidates down gently," Carlisle said. The largest part of a Google "interview" is work sampling. An engineer will have to code an app; a public relations person will be asked to write a press release.

Google has expended much effort trying to gauge how much weight to place on interviews relative to everything else. Another important question is how many interviews to schedule for a given candidate. More is better, up to a point—but "we don't want to waste people's time," Carlisle said. His statistical analysis found that about five interviews was optimal. Interviewing any more had diminishing returns.

In 2003 the Corporate Executive Board did a similar but broader study, asking twenty-eight thousand recently hired employees nationwide how many interviews they went through to get their job and comparing the answers to subsequent performance reviews. The results were similar to Carlisle's: the best workers were those hired after four or five interviews.

It might be that the employees interviewed more than five times tended to be those who sent mixed signals. Some of the interviews went well, some didn't, and the company scheduled more than the usual number to resolve the conflict. Those doubts were justified, the survey results imply.

An alternate explanation for the Corporate Executive Board results is that the best people got fed up with too many interviews and said, "Enough!" The companies that insisted on eight or ten interviews got stuck with the truly needy.

A Google candidate can expect about five on-site interviews, by five different interviewers, scheduled in a single day. One is a lunch interview, in which the candidate is supposed to be granted sufficient relief from tough questions to sample the gourmet cuisine. Interviewers give candidates one of four "grades." According to Carlisle, they mean "I don't think we should hire this candidate"; "I don't think we should hire this candidate but I can be convinced otherwise"; "I think we should hire but I can be convinced otherwise"; and "a strong 'hire.' "

Google's interviewers do not directly make hiring decisions. Their job is to conduct good, tough interviews and report the results. The reports explain what questions were asked, what the answers were, and what the interviewer thought of the answers. "Wisdom of crowds" (one of the principles behind Google's hiring) works best when each judge is allowed to form an opinion independently of everyone else. Then an average of the opinions is likely to be close to the truth. At Google, interviewers are told not to discuss the candidate with one another until they've submitted their report.

(The one legitimate reason for passing notes is to avoid wasting everyone's time when a candidate is clearly unsuitable. Google has interviewers report briefly to the candidate's recruiter, who has the power to halt the interviews in those rare unhappy cases.)

The interviewers' reports become part of the candidate's package, distributed to a hiring committee. If the committee approves the candidate, another committee reviews the package, and then another. Finally, every hiring decision still goes by Larry Page for approval. At Google, hiring is more bureaucracy than algorithm.

That's the polar opposite of what most people probably expect. "Ultimately what we are trying to do is to make the process fair and to eliminate as many biases as possible," said Setty. Google takes the problem of "biases" seriously, and here it's not just ethnic or gender bias (which are taken seriously indeed) but bias in the broad sense of any habitual quirk in decision making. For instance, an interviewer may have a rule of thumb that "you can't go wrong hiring a Stanford PhD." If that's placing too much weight on one datum, as it probably is, that would be a bias. It would equally be a bias if an employer insisted that degrees and universities don't matter and gave them no weight at all. Google's nominal aim is to assign optimum weight to everything. That may be an unreachable goal, but it guides the thinking of Google's People Ops.

The interviewers assigned a candidate are chosen to reflect a variety of backgrounds, personalities, genders, ages, and ethnicities. This practice acknowledges that it's human nature to relate more easily to candidates from familiar universities, with familiar life histories, with familiar styles of dress or speech. "We're not trying to take the human element out of this at all," Setty said. The goal is "understanding what these patterns are and then presenting [them] to the people who are making the decision, not making the decision for them." "In most companies, you as a manager go to finance and say, 'Do I have the budget to hire a person?' And then you go off, you have your own system, you talk to people you know, you just go and hire them. Here, it's the extreme opposite of that. We say, 'You don't get to make a decision because we think that everyone has their own biases and we can't trust any one individual to make the right decision.' So we have this seemingly very complex and time-consuming process. But ultimately the goal is to reduce false positives."

False Positives and False Negatives

A *false positive* is the outcome when a candidate passes the vetting process and is hired, only to be a poor employee. The opposite is a *false negative,* when a candidate who would have been a good employee is rejected. It might seem that false positives and false negatives are equally bad. However, they're not equal at Google, or anywhere else.

Job seekers fear false negatives because they mean bad things happen to good people—a tough interviewer or a bungled question prevents you from getting a job where you could have excelled. From the applicant's perspective, that is deeply unfair. But to the employer, false negatives are invisible. "We don't know whether our system has a lot of false negatives that we screen out of the process," admitted Setty. "We don't know that because we haven't hired them."

Bad hires, on the other hand, are in everyone's face. Reducing false positives is People Ops' core directive, "the belief of Larry, Sergey, and Eric, since the founding of the organization," said Setty. That is why Google's hiring process is a paragon of redundancy.

This isn't just a Google thing. "In an up market, say the late 1990s, the cost of making a bad hiring decision was low," according to Alec Levenson of the University of Southern California's Center for Effective Organizations. "The company could be a lot more cavalier about hiring, because if the worker doesn't fit, the chances are that he'll move on soon."

Not now. Employees cling to jobs like limpets to wet rocks. The more marginal the employee, the stronger his suction. The only way to get rid of a questionable hire is to fire him. That's a fraught process. "There's been gradual erosion over the past 30 years of pure employment-at-will as more and more people have come under employment protection laws" explains Levenson. "It's become more and more difficult for companies to... hire and fire. Even if 100 people are eligible to sue, only one or two might, but that's all it takes." Hiring today is like marriage used to be: for the duration.

Despite the concern with false positives, "you can totally bomb an interview," Carlisle insists, "and that's not the end of your candidacy." Google is aware that interviews are a noisy signal. There is some evidence that candidates who get an enthusiastic thumbs-up from one interviewer perform better, on the average, than candidates who get merely favorable grades from all the interviewers. It's like the candidate is an indie film: better to inspire passion in somebody than to try to please everybody. The flip side of this is that getting a single poor review isn't so bad.

The Obama Question

Google's interviewers are discouraged from asking the traditional brainteasers popular at other companies, like "Why are manhole covers round?" They're also not supposed to test candidates' knowledge with trivia questions like the following:

? Explain the significance of "dead beef."

Nor are they supposed to confuse interviewees with cryptic demands like this one.

? There's a latency problem in South Africa. Diagnose it.

The case against these questions is that their pat answers aren't informative and are too easily remembered. But engineers at Google, like those everywhere else, only half pay attention to what HR people say and ask these questions anyway. (In case you've missed it, the answer to the manhole question is "because a round manhole cover, unlike a square one, can't fall in the hole." The answers to "dead beef" and "latency problem" are on pp. 169–171.)

"You're supposed to ask open-ended questions that test problem solving and general knowledge, then get into specifics," explained one former Google interviewer. Google's most characteristic, and most emulated, interview questions are short questions that spark conversations.

On January 26, 2008, senator and presidential hopeful Barack Obama attempted to establish his new-economy credentials. He visited the Googleplex for a public chat with Eric Schmidt. Schmidt commented that it was hard to get a job as president—and hard to get a job at Google. In order to test Obama's qualifications, Schmidt asked him, "What's the most efficient way to sort a million 32-bit integers?"

Obama's reply was, "The bubble sort would be the wrong way to go."

It was scripted shtick, of course, and it got a big laugh. The video is on YouTube.

The "Obama question" is asked quite seriously of software engineer candidates at Google. Such people know that bubble sort is miserably slow (hence Obama's punch line). Using bubble sort for a million numbers would be like filling a swimming pool using a thimble.

The best answer to this question begins, "It depends." It depends on the makeup of the list of integers and the constraints of time and memory. The applicant is expected to ask about these things. The question is intended to lead to a discussion of the relative merits of algorithms and how the applicant would go about choosing the best tool for the job (perhaps the hardest thing of all to teach, or to learn).

"In general, we're not trying to fill a particular job," explained Setty. "The way Google has morphed and grown, we find that people can join in a particular role, and five years later they're doing something completely opposite. You cannot just hire for a job; we want to hire for Google as a whole."

This makes it useful to ask some questions not tied to a specific set of skills. Google's broader, more playful questions—like "thrown into a blender"—encourage the candidate to engage with the interviewer and devise a coherent solution. Google's interviewers are like good journalists in that they keep asking follow-up questions designed to exclude pat answers. A running theme is, *can you improve that answer further?*

Does Your Facebook Page Count?

Much of what Google does is redefining the notion of privacy. It's become the stuff of urban legends. Did you hear about the woman who got the goods on her cheating husband? She spotted his 4x4 in front of his mistress's home on Google Maps' Street View—then Googled a good divorce lawyer. (Reported as fact by *The Sun* in 2009, it's since been debunked.) Another tall tale is that Google checks its candidates' search histories based on their computer's IP address. This would tell Google what other companies' employment sites you've visited recently—uh, as well as a lot of other interesting stuff.

While that is a canard, Google, along with companies everywhere, is wrestling with the proper role of social network sites in hiring. Is it legitimate for an employer to use information on Facebook, YouTube, or Twitter in a hiring decision? It's an interesting ethical poser, but honestly, that ship has sailed.

In 2007, the cofounder of LinkedIn, Reid Hoffman, was looking for a new CEO for the company. He questioned the traditional importance of references in vetting executives. "It's a low bar for someone to give you two or three people who'll say nice things about them," Hoffman observed. Instead, he used the LinkedIn network to assemble a list of twenty-three business contacts of the lead candidate—whom the candidate *hadn't* listed as references. Some were two degrees of separation removed. These "off-balance references" were contacted for a more unfiltered take. "Our way of doing it requires a bit of detective work, and you need to put a story together," Hoffman said. "But you quickly sense if a person is good or a sham."

What was edgy in 2007 is now mainstream. A CareerBuilder survey found that the proportion of employers checking out candidates on social network sites surged from 22 percent in 2008 to 45 percent in 2009. It's safe to assume the number will be higher by the time you read this.

"People are willing to tell you all sorts of stuff about themselves on Facebook and LinkedIn," said Google's Todd Carlisle. That's compelling in a corporate culture that believes it's a sin to ignore information about a potential hire. "What we have to balance is, they're not sending it to Google and they're not applying for a job. We want to be cautious of that."

Not everyone is so cautious. In the 2009 survey (of 2,667 managers and HR people), 35 percent said they decided not to offer a job to someone based on what they found on Facebook, Myspace, or other sites. The biggest red flags, according to employers, were "provocative or inappropriate photographs or information" (53 percent) and "content about... drinking or using drugs" (43 percent). In other words: just about what you'd expect to find on a Facebook page.

Applicants are also Googling their interviewers. "I always try to get a list of people I'll be interviewing with beforehand and Google them, see if they're on Twitter, see if they have a blog," said Rakesh Agrawal, who's been on both sides of the interview table. "Anyone who makes that extra bit of effort certainly gets a few extra points." When he's the interviewer, Agrawal poses questions based on what he's learned online. "The trick there is just to do that without being creepy. You don't want to have stalked the person, to have gone through their Flickr photos for some obscure picture from ten years ago."

Aside from privacy concerns, there's the possibility of applicants' gaming the social networks, said Carlisle. A candidate who believes a potential employer is going to covertly look at his Facebook page could add fake accomplishments or have friends do so. A false credential on a CV is a dismissible offense. On a social page, it's still uncharted territory.

So far the missing ingredient is mutual disclosure. Employers need to admit they're looking at social network sites. Job hunters need to be able to vouch for career-relevant information on those sites.

As the amount of personal information on the web grows exponentially, "it gets harder and harder to crawl all the information about somebody," said Carlisle. "So something that pieces together what's on LinkedIn, what's on Facebook, what's in their YouTube videos, what did they put on their application, and combines it all together in an easier way, that would be compelling for me." It's a pipe dream so far, but it's likely to be here soon.

In the meantime, it's best to follow the standard advice for job seekers. Set social network pages to "private" and/or tidy them up before a job search. One finding of the CareerBuilder survey is particularly sobering. It was the importance placed on "poor communication skills." Twenty-nine percent of employers cited that as a potential deal breaker, and it covered a multitude of grammar-police offenses. Sixteen percent chose not to consider a candidate because his application or e-mail correspondence used "text language" (like "GR8").

"Not hiring someone for poor communication skills on Facebook is a bit silly," one observer said, "like saying I won't hire you because you told a joke with bad grammar at the restaurant last week, when I was in the booth behind you spying."

Well said—but you'd better watch spelling and grammar nonetheless.

QUESTIONS
Classic Google Riddles

Here are some of the more offbeat questions that Google interviewers are known for asking. Only one ("describe a chicken") demands a computer background, but all are challenging, and most have been adopted at other companies. (Answers begin on p. 171.)

? Design an evacuation plan for San Francisco.

? Imagine a country where all the parents want to have a boy. Every family keeps having children until they have a boy;

then they stop. What is the proportion of boys to girls in this country?

- **?** On a deserted road, the probability of observing a car during a thirty-minute period is 95 percent. What is the chance of observing a car in a ten-minute period?
- **?** You have a choice of two wagers: One, you're given a basketball and have one chance to sink it for £1,000. Two, you have to make two out of three shots, for the same £1,000. Which do you prefer?
- **?** Use a programming language to describe a chicken.
- **?** There's a staircase and you're allowed to ascend one or two steps at a time. How many ways are there to reach the *N*th step?
- **?** You have *N* companies and want to merge them into one big company. How many different ways are there to do it?
- **?** What is the most beautiful equation you have ever seen? Explain.

Five

Engineers and How Not to Think Like Them

The Value of Keeping Things Simple

The great physicist Richard Feynman once applied for a job at Microsoft (so runs the guaranteed-apocryphal story). "Well, well, Dr. Feynman," the interviewer began. "We don't get many Nobel Prize winners, even at Microsoft! But before we can hire you, there's a slight formality. We need to ask you a question to test your creative reasoning ability. The question is, why are manhole covers round?"

"That's a ridiculous question," Feynman said. "For one thing, not all covers are round. Some are square!"

"But considering just the round ones, now," the interviewer went on, "why are they round?"

"Why are round manhole covers round?! Round covers are round by definition! It's a tautology."

"Uh—right. If you'll excuse me, Dr. Feynman, I'd like to consult with our human resources department." The interviewer left the room for ten minutes. When he returned, he announced, "I'm happy to say that we're recommending you for immediate hiring into our marketing department."

This joke pokes fun at one of the most famous brainteaser questions, long associated with Microsoft and alleged to have been devised by Steve Ballmer himself. It expresses deep ambivalence about this style of interviewing. Feynman (a childhood hero of Sergey Brin's) shows more creative thinking than does Microsoft's so-called right answer.

A true story: Brin did graduate work in Stanford's computer science building, named for its donor, William Gates. Each room in the Gates Building had a four-digit number. "We were offended at having four-digit numbers when you don't have ten thousand rooms," Brin explained. He devised a new numbering system using three digits. The building doesn't have a thousand rooms, either, but Brin reasoned that he had to retain the convention that the first digit gives the floor. "I just had the numbers roll around the building," Brin said. "Even numbers were exterior, odd numbers were interior. The second digit told you how far around the building you had to go."

Google's people like to think of themselves as having a uniquely creative approach to design. In this worldview, Microsoft is sometimes cast as a bad example. Though the invidious comparisons have less to do with the real Microsoft than with outsiders' jokes, there is a slender basis in history. Microsoft began in an age when small computers were for hobbyists and everyone wrote spaghetti code. Google was founded a generation later, when a new discipline of algorithm theory had changed the way software was written. Microsoft has, of course, recruited many of the world's best coders and computer scientists. But it has baggage—legacy products, legacy users, and a corporate culture forged in the 1980s. Google entered the new millennium with a clean slate. As the tech blogger Joel Spolsky wrote,

> A very senior Microsoft developer who moved to Google told me that Google works and thinks at a higher level of abstraction than Microsoft. "Google uses Bayesian filtering the way Microsoft uses the 'if' statement," he said. That's

true. Google also uses full-text-search-of-the-entire-Internet the way Microsoft uses little tables that list what error IDs correspond to which help text. Look at how Google does spell checking: it's not based on dictionaries; it's based on word usage statistics of the entire Internet, which is why Google knows how to correct my name, misspelled, and Microsoft Word doesn't.

Bob and Eve

This "higher level of abstraction" figures in many of Google's interview questions. Try this one:

? You want to make sure that Bob has your phone number. You can't ask him directly. Instead you have to write a message to him on a card and hand it to Eve, who will act as go-between. Eve will give the card to Bob, and he will hand his message to Eve, who will hand it to you. You don't want Eve to learn your phone number. What do you ask Bob?

This question is usually asked of software engineers, who instantly recognize the names "Bob" and "Eve." In computer science textbooks, it's conventional to speak of "Alice" sending a coded message to "Bob" (it sounds a bit more human than saying "A sends a message to B"). The rote villain of the textbooks is a snoop called "Eve" (for "eavesdropper"). Coded messages are vitally important on the Internet—they're the basis of e-commerce and cloud computing. Eve's many guises include hackers, spammers, and phishers. It is not too much of an exaggeration to say that this interview question presents, in a kernel, the central problem of our wired world.

It also reveals very different ways of thinking about problems. There is a technically brilliant solution. Every textbook discussion of Bob and Eve segues into an exposition of RSA cryptography, the type used by PayPal and other forms of electronic commerce. Suffice it to say that RSA involves some heavy computation. That's okay, as it's always done by computer. The smart interviewee is led to wonder whether there's any way to explain RSA to Bob, as part of a message that could fit on the back of a business card. This is something like telling your grandmother how to build an iPad so clearly that she could make one.

It can be done! It's possible to explain a bare-bones implementation of RSA to a naive Bob who doesn't know how to code. (I give the whole explanation in the "Answers" section.) A lean version of the instructions will fit on a three-by-five-inch index card, even a business card if you've got microscopic handwriting. The candidate who succeeds in drafting his RSA message to Bob will feel he's knocked the ball out of the park.

Not so fast. He's just given the "Microsoft answer." Eve or no Eve, Bob is sure to balk at following complicated instructions just for the mundane task of confirming a phone number. Google interviewers expect engineers to know RSA, of course, but they're especially impressed by those who come up with a simpler, more practical answer.

Bob, I need you to follow these instructions carefully without questioning them. Pretend my phone number is a regular 10-digit number. First, I need you to cube the number (multiply it by itself, and then multiply the product by the original number again). The answer, which will be a 30-digit number, has to be exact. Do it by hand if you have to, and double-check it. Then I need you to do the longest long division you've ever done. Divide the result by this number: 50533669377341834823. The division also has to be exact. Send me the remainder of the division only. It's important that you don't send the whole part of the quotient – just the remainder.

The Microsoft Answer

Bob, call my phone using the number you've got.

The Google Answer

Tell Bob to call you (ideally, give a specific time). If your phone rings, bingo. If not, that tells you he doesn't have the right number. That's all the question asks for ("You want to make sure that Bob has your phone number..."). Why do things the hard way?

This question tests something rarer than education—the capacity to ignore what you learned when it isn't helpful. In business there's no one to tell you what part of your education applies (if any). There is an overwhelming temptation to use whatever intellectual tools are at hand and to pat yourself on the back for using the highest-powered tool possible. Google doesn't want people who instinctively do things the hard way because they can. They want those with a knack for intuiting simple solutions that work.

Considering the Human Element

This raises the all-important question, what separates the entrepreneur from the engineer? In part, it's the ability to not think like an engineer sometimes. An engineer can't help falling in love with the clever ideas and algorithms that have gone into a new product. An entrepreneur has to ignore them and judge whether the end users will want, or be able, to use the product. Since Google is a place where job descriptions are fluid, it tries to find employees with the ability to put themselves in someone else's head. Many of its interview questions bear on that theme.

? An executioner lines up 100 prisoners single file and puts a red or a blue hat on each prisoner's head. Every prisoner can see the hats of the people in front of him in line—but not his own hat, nor those of anyone behind him. The executioner starts at the end of the line and asks the last prisoner the color of his hat. He must answer "red" or "blue." If he answers correctly, he is allowed to live. If he gives the wrong answer, he is killed instantly and silently. (While everyone hears the answers, no one knows whether an answer was right.) On the night before a lineup, the prisoners confer on a strategy to help them. What should they do?

Like "Bob and Eve," this feints at being a familiar type of question, in this case an old-school logic puzzle. It's unquestionably related to a riddle created by the American mathematician and computer scientist Alonzo Church (1903–95). In the 1930s, Church invented a puzzle about three gardeners who have spots of dirt on their foreheads. No one can see his own forehead, of course, and there's no mirror. The gardeners are told that at least one has dirt on his forehead and have to deduce who is smudged. One of Church's students, the logician Raymond Smullyan, took the idea and ran with it, producing scores of ingenious puzzles in a series of popular books. In later variations by Smullyan and others, the telltale "spots of dirt" are replaced with hats or colored dots or anything that might plausibly be visible to everyone except the wearer. Stories are embellished with amusing motives, like trying to deduce whose wife is cheating on whom, or who is the spy. Many of these puzzles are used in interviews throughout the corporate world.

Solutions usually hinge on the assumption that everyone involved is a "perfect logician." That means that person A can deduce *x* based on what B deduced from *y* and C failed to deduce from *z*. It will not come as a shock to learn that the intended solutions have nothing to do with the real world. In many cases, these puzzles penalize those who understand how real people think and act.

As it's used at Google, the executioner question is a subtle deconstruction of the genre. There is no definitive right answer. The best responses show appreciation of the human element and its unintended consequences.

Start by looking at what happens when the prisoners have no plan at all. Should the prisoners just guess their hat colors randomly, they would be right about 50 percent of the time. That means that about 50 of the 100 would survive, on average.

Any plan must do better than that to be worthwhile. Conceptually, the prisoners want to send information up the line. The last prisoner, #100, sees the color of everyone's hat except his own. Were he free to speak his last words, he could recite the hat colors of the 99 prisoners in front of him, saving them all. That's forbidden. He is permitted only one word, and it must be either "red" or "blue." That's a one-bit message when he really wants to send a ninety-nine-bit message.

Because this last prisoner has no way of learning his own hat's color, he has nothing to lose. He might as well use his answer to do something useful, like name the hat color of the guy directly in front of him. This will permit #99 to give the right answer. Number one hundred will have a fifty-fifty chance of being spared, too, since his hat might happen to be the same color as #99's.

Why not have everyone answer with the hat color of the prisoner in front? That won't work. Imagine you're somewhere in the middle of the line. The prisoner behind you says "red," meaning *your* hat is red. The hat in front of you is blue. Do you save your own life by answering "red," or do you pass on the right answer to the prisoner ahead of you by saying "blue"? You can't do both.

One scheme—call it plan A—is for the even-numbered prisoners to answer with the hat color they see immediately in front of them, and for the (lucky) odd-numbered people to take advantage of this information to save themselves. Under plan A, all 50 odd-numbered people will live. The even-numbered people have to take their chances. You can expect that about half of them will be killed. In all, this brings the number of likely survivors up to 75. That's certainly better than not having a plan at all.

This survival rate could be improved, were the prisoners able to send poker-cheat "signals." Let prisoner #100 name the color of #99's hat. Number ninety-nine clears his throat before answering if and only if the hat in front is the same color as his own. He then names his correct hat color. The throat-clearing signal would allow #98 to name his hat color and send a similar signal to #97. Everyone except #100 could save himself, and #100 would still have a fifty-fifty chance of survival. That's a survival rate of 99.5 out of 100.

Secret signals are not usually considered a legitimate solution. You're to assume that each prisoner's answer must be "red" or "blue," and nothing more. There is an entirely legal scheme that's about as good. Plan B, let's call it, has the last prisoner count the number of red hats he sees in front of him and answer according to whether this number is odd or even. The rule could be, "red" means "the number of red hats I see is odd," and "blue" means "the number of red hats I see is even."

Plan B doesn't help the last prisoner's chances of survival (nothing can). It does offer a way to save everyone else. Say you're #99 and you hear #100 say "blue." That means he sees an even number of red hats. You count the number of red hats *you* see in front of you. Is it also even? If so, then your hat cannot be one of the red hats #100 saw. Therefore, your own hat must be blue. By saying "blue," you save your own skin. The beauty of the plan is that "blue" also helps #98. He knows that #100 saw an even number of red hats and that yours wasn't one of them (because you gave the same answer). This allows #98 to deduce his own hat's color.

In general, everyone knows the parity of the total number of red hats (exclusive of prisoner #100's, which is out of sight and out of mind). Everyone also knows the colors of the hats behind him (since this has been announced) and can use that to deduce his own hat's color. By announcing the color of his own hat, each prisoner saves himself and supplies the information subsequent prisoners will need.

Plan B requires that everyone remember #100's answer and keep a running total of how many people behind them said "red." Each needs to add this number to the number of red hats he sees in front of him, then compare the sum with #100's answer. If #100 said "red," meaning he saw an odd number of red hats—and if you're aware of 47 red hats (the 21 you see and the 26 people behind you you've heard say "red")—then that jibes. All the reds are accounted for, and your hat must be blue. If there is a discrepancy, it must be due to your hat's being red. You say whatever color your hat is.

Does the above paragraph strike you as a little confusing? Now imagine you're explaining it to a hundred prisoners, or to your sister-in-law, or to the Midwest sales force. Real people make mistakes, especially when someone's about to garrote them. Let just one prisoner slip up, and the plan falls apart.

There are some engineers who wouldn't think of this. They are happy to stop with a technically valid answer that no one can understand. That's why your TV setup has four remotes, all confusing.

The better answers to this question go beyond the mere logic-puzzle response. A candidate should ask, how practical is plan B, really?

One saving grace is that there are only two possible answers. When somebody gets confused and screws up, there's still an even chance he'll say the right thing by dumb luck. But in the long run, the inevitable errors compromise plan B. Remember, Google likes answers that scale up. For "100 prisoners" read "1,000" or "an indefinite large number." As John Maynard Keynes said, in the long run, we're all dead. Well, close. When the line is long enough that some make errors, about half the prisoners will be acting on the right information and surviving—while the other half will be acting on wrong information and be slaughtered like sitting ducks.

Allowing for errors, the asymptotic survival rate of plan B is only 50 percent. That's no better than what you'd expect by having no plan at all.

At any rate, plan B's real survival rate may be less than the 75 percent rate offered by plan A (where errors do not propagate forward). Not that plan A is foolproof, either. The even-numbered prisoners are asked to nobly save the odd-numbered ones and get nothing in return. Picture how that would go down at Belmarsh. However it's decided who gets to be the odd-numbered ones, the division will stoke long-simmering resentments. The even-numbered prisoners can't improve their own fifty-fifty survival chance. But a malicious prisoner could settle a score by intentionally giving the wrong answer to an odd-numbered prisoner in front of him. Since errors don't propagate, the realistic survival rate of plan A is only a little less than the theoretical 75 percent.

Ideally the prisoners should do trial runs of both plans and choose the one with the higher survival rate—in practice. It's good to mention that to conclude your answer. In this and more realistic situations, no one can predict infallibly how people will take to a new idea—you have to try it out.

Listen to Your Mother

Who's smarter, a computer science PhD or your mum? After years of interviewing at Google, Paul Tyma, an engineer, resolved to find out. Suppose you're given a million sheets of paper (runs one of Tyma's interview riddles). Each is the record of a university student. You're to sort them in order of age (number of years old). How would you do it?

Tyma posed this question to his mother, who knew absolutely nothing about computer science. Mrs. Tyma's answer was more efficient than those of many of the highly educated applicants Tyma has interviewed.

How can this be? Mrs. Tyma's answer was, she'd make stacks. Take the first record off the top of the pile and look at the age. If it's a twenty-one-year-old, it goes in the twenty-one-year-olds' stack. If the next record is a nineteen-year-old, it goes in the nineteen-year-olds' stack. And so on. You have to look at each record only once, and when you're finished, you simply collect the stacks in order of ascending age. Done!

That procedure is about twenty times faster than quicksort, the algorithm that many Google applicants suggest. Some candidates go nuts when told of Mrs. Tyma's solution. Quicksort is "guaranteed" to be asymptotically fastest! The textbook said so!

They're forgetting the mathematical fine print. Quicksort is based on comparisons: is this number bigger than that? You don't always need comparisons to sort. You don't here because there are many, many records and only a few distinct ages of students. Quicksort is versatile, like a Swiss Army knife, but Mum's stacks happen to be a much better tool for this particular task.

One of the oft-cited mysteries of creativity is that revolutionary ideas often come from nonexperts with an outsider's perspective. The computer science graduates were so used to thinking of high-powered algorithms that they couldn't *not* think of them. Without that mental baggage, Mrs. Tyma intuited a better solution. Sometimes "creativity" is just common sense.

With apologies to the folks in Redmond, I'll end on another Microsoft joke because it makes the point well (a point that applies everywhere, not just at Microsoft): A helicopter was flying around above Seattle when a malfunction disabled all of its electronic navigation and communications equipment. The clouds were so thick that the pilot couldn't tell where he was. Finally, the pilot saw a tall building, flew toward it, circled, and held up a handwritten sign that said WHERE AM I? in large letters. People in the tall building quickly responded to the aircraft, drawing their own large sign: YOU ARE IN A HELICOPTER. The pilot smiled, looked at his map, determined the route to Sea-Tac Airport, and landed safely. After they were on the ground, the copilot asked the pilot how he had done it. "I knew it had to be the Microsoft building," he said, "because they gave me a technically correct but completely useless answer."

QUESTIONS
Avoiding Technically Correct but Useless Answers

Each of these questions has a simple, practical answer and a complicated or useless one. That's a hint, but be warned: it's often easier to find the complicated answer than the simple one. (Answers begin on p. 193.)

- **?** If you had a stack of pennies as tall as the Empire State Building, could you fit them all in one room?
- **?** You have ten thousand Apache servers, and one day to generate £1 million. What do you do?
- **?** There are two rabbits, Speedy and Sluggo. When they run a 100-meter race, Speedy crosses the finish line while Sluggo is at the 90-meter mark. (Both rabbits run at a constant speed.) Now we match them up in a handicapped race. Speedy has to start from 10 meters behind the start (and run 110 meters), while Sluggo starts at the usual mark and runs 100 meters. Who will win?
- **?** You've got an analog watch with a second hand. How many times a day do all three of the watch's hands overlap?
- **?** You're playing football on a desert island and want to toss a coin to decide the advantage. Unfortunately, the only coin on the island is bent and is seriously biased. How can you use the biased coin to make a fair decision?

Six

A Field Guide to Devious Interview Questions

Decoding the Interviewer's Hidden Agendas

Interview advice abounds, not all of it useful. One semi-myth of today's competitive interviews is that the interviewer is there to help you. Interviewers are out to find the most qualified person for each opening. Ten-to-one odds (or steeper) say you're not it.

The myth of the benevolent interviewer is perpetuated by lists of pointers on employment websites. Google sends job candidates this advice in an e-mail:

> At Google, we believe in collaboration and sharing ideas. Most importantly, you'll need more information from the interviewer to analyze & answer the question to its full extent.
>
> - It's OK to question your interviewer.
> - When asked to provide a solution, first define and framework the problem as you see it.
> - If you don't understand—ask for help or clarification.
> - If you need to assume something—verbally check it's a correct assumption!

- Describe how you want to tackle solving each part of the question.
- Always let your interviewer know what you are thinking as he/she will be as interested in your process of thought as your solution. Also, if you're stuck, they may provide hints if they know what you're doing.
- Finally, listen—don't miss a hint if your interviewer is trying to assist you!

Don't get me wrong—everything here is good advice as far as it goes. Just don't expect the interviewer to "collaborate" by answering the question for you. They're also not telling how positively unnerving the "numb," poker-faced demeanor of Google interviewers can be. As one Google interviewee said, "You will have this 'lost in space feeling' when you do not know if you are saying interesting or stupid things."

The poker face is partly a matter of fairness. One thing interviewers routinely fail to appreciate is how subjective their assessments of answers are. This is true even when presenting tasks with unique "right answers." Inevitably, it's the interviewer's prerogative to call, "time!" by cutting a struggling candidate off and moving on. (Who knows whether the candidate would have gotten the answer had he been given a few more seconds?) Interviewers may unconsciously give more time to candidates they like, less to those they don't. Google's interviewers are therefore encouraged to aim for consistency. They try not to give hints through tone of voice or body language. When hints are offered, they should be offered in the same way for every candidate (e.g., "If she asks whether the blender has a lid, say no; otherwise don't mention a lid"). Don't expect much hand-holding.

Video games fall into genres like first-person shooter, strategy, and life simulation. A player who failed to understand the genre conventions would be at a disadvantage. The same goes for tricky interview questions, which can also be grouped into more or less distinct categories. It's important to recognize the type of question being asked, as this is your first clue about how to proceed. Here's a quick guide to the major types of mental-challenge questions and how to answer them.

Classic Logic Puzzles

These questions are often very old—some of them can literally be traced back to the Middle Ages. They are featured in articles, books, and video games. Here's an example:

> Two Cambridge graduates meet for the first time in 20 years.
> A asks, "How've you been?"
> B says, "Great! I'm married now and have three daughters."
> A asks, "How old are they?"
> B says, "The product of their ages is 72, and the sum of their ages is the same as the number on that building over there."
> A says, "Uh... I still don't know."
> B says, "My oldest just started to play the piano."
> Then A says, "Really? My oldest is the same age!"
> How old are the daughters?

Solving a puzzle like this is mostly a matter of persistence. Start from the top. It's news to A that B is married with kids. They haven't seen each other in 20 years, so all B's daughters must be 20 years old or less. You can probably take 19 as the maximum.

Try some algebra, for what it's worth. Call the daughters' ages x, y, and z. The product is 72.

$$xyz = 72$$

The second part of B's comment is more puzzling. Rather than telling the sum of the ages, he says it's the same as the number on a building. We don't learn the number, though.

$$x + y + z = \text{number on building}$$

Does that tell anything at all? It's hard to see how. We're not even told that it's a street number. But let's say it is. The only usual restrictions on street numbers are that they aren't negative or zero (or irrational or imaginary...). Some street numbers are fractional. But we can probably rule out fractions. Ages are usually given as whole numbers. Indeed, A says he can't deduce the ages from the above information.

Still, the bit about the number on a building is likely to be significant. Logic puzzles are like poems or code: the good ones don't contain the inessential. The fact that the question talks about the number on the building must mean something—we just don't know what.

B says his oldest one started to play the piano. This means that there is a daughter whom B casually refers to as "the oldest." He wouldn't likely say that if the piano-playing daughter had a twin sister—not even if she was born five minutes earlier and was technically "the oldest." Likewise, a father wouldn't refer to a triplet as "the oldest."

Therefore the three daughters are *not* triplets and the oldest two are *not* twins. Anything more? Well, it takes about nine months to make a baby. Sisters who are not twins are usually of different ages, meaning at least one year apart. This is not an ironclad conclusion. Two sisters could have been born eleven months apart and be the "same" age one month out of every year. The two Cambridge graduates could have met during that month.

But let's give it a try. Assume until proven otherwise that

- the three daughters are not all the same age; and
- the two oldest are not the same age.

After B's statement about the "oldest," the lightbulb goes on above A's head. He concludes that his oldest is the same age.

How can that be? It's time to go back and look at the one complete equation we've got: $xyz = 72$. Since ages are whole numbers, there are only so many combinations that will work. Seventy-two is 6×12, and its prime factors are $2 \times 3 \times 3 \times 4$. We can also factor out a 1 (as in $1 \times 1 \times 72 = 72$). Here, then, is the complete list of whole-number triplets whose product is 72. In a job interview, you'd write this on the whiteboard:

$1 \times 1 \times 72$

$1 \times 2 \times 36$

$1 \times 4 \times 18$

$2 \times 2 \times 18$

$2 \times 3 \times 12$

$1 \times 6 \times 12$

$1 \times 8 \times 9$

$2 \times 4 \times 9$

$3 \times 3 \times 8$

$2 \times 6 \times 6$

$3 \times 4 \times 6$

We can rule out the first two. All the daughters have to be less than 20. We can probably rule out 2, 6, and 6, too. But remember, B didn't mention "the oldest" until the end, and most of A's reasoning was done without the benefit of that comment. Just for fun, let's take the sums of the ages. There has to be a reason why the story mentions a number on a building.

1 + 1 + 72 = 74

1 + 2 + 36 = 39

1 + 4 + 18 = 23

2 + 2 + 18 = 22

2 + 3 + 12 = 17

1 + 6 + 12 = 19

1 + 8 + 9 = 18

2 + 4 + 9 = 15

3 + 3 + 8 = 14

2 + 6 + 6 = 14

3 + 4 + 6 = 13

Thirteen! Three and 4 and 6 add up to an "unlucky" number that is unlikely to be a street number in our superstitious world. Again, this isn't a foolproof deduction. Thirteen *is* sometimes used as a street number, and there are other reasons why "13" might appear on a building (such as an ad for TV channel 13). But this is exactly the kind of twist you should be looking for in a logic puzzle. You're expecting a mathematical deduction, so they toss in a cultural one.

Rule out 3, 4, and 6 because their sum, 13, probably wouldn't be on a building. This "brilliant" deduction gets us exactly nowhere (which is also typical of these puzzles). There remain at least seven plausible sets of ages whose products are 72 and whose sum is not a jinx.

Wait—the guys in the story know something we don't, namely, the number on the building. It's visible to both, and that's why they don't bother to mention it. Despite that, A is baffled after the comment about the sum of the ages equaling the number on the building. Since Cambridge graduates never miss a trick (a somewhat iffy assumption), it appears that the building number does not itself give enough information to solve the puzzle. The only way that's possible is for the number on the building to be 14. There are *two* sets of ages that sum to 14. Had the building number been 18, for instance, A would know the ages had to be 1, 8, and 9.

The two possible sets of ages are

3, 3, 8

2, 6, 6

When B said his oldest is starting piano, that nailed it. It rules out (2, 6, 6) where the two oldest daughters are the same age. That means the ages of B's daughters must be 3, 3, and 8. That is the one and only correct answer. The solution is arrived at through a lot of back-and-forth reasoning rather than a single burst of insight. It's like untangling the cords behind a desk.

In his longtime *Scientific American* column, Martin Gardner published a variant of this puzzle, crediting Mel Stover of Winnipeg as the first of several readers to send it in. Gardner did not otherwise know the origin, suggesting it may have been new at the time. In that version, the product of the ages was 36, the sum equaled an unspecified house number, and the parent mentions that the oldest child has a wart on his left thumb.

When the product of the ages is 36, there are fewer combinations to consider. The house number *has* to be 13, as that's the only sum permitting two solutions. (The correct answer is 2, 2, and 9.) Perhaps someone realized that 13 is an unlikely house number and adjusted the puzzle accordingly.

You're probably thinking, "Great! I now know how to answer that question, should it ever come up. What if I'm asked something else?"

To a limited extent, there is a pattern. Like jokes or golf courses or haiku, logic puzzles aspire to a certain form of cleverness and play by certain rules to achieve it. By reverse engineering this structure, you can come up with a three-part process that applies, in broad outline, to the solution of most of these puzzles. It goes like this:

1. Distrust the first answer or line of attack that pops into your head. It won't work because if it did, the puzzle would be too easy.
2. Decide what feature of the question's wording doesn't "fit" and take that as a clue.
3. Look for a solution that's surprising in some way.

The knee-jerk reaction to the Cambridge graduates puzzle is that this is a "story problem" to be translated into algebraic equations and solved. As we've seen, anyone who narrowly insists on algebra will end up hitting his head against a wall.

B makes two statements that don't "fit." He refers to a number on a building, but we're not told what that number is. His mention of his oldest daughter's playing the piano is a complete non sequitur—yet it's this that allows A, and ultimately the puzzle's audience, to solve the mystery.

The third step may be the most important of all. Good puzzles feature an element of surprise. Either the answer itself is surprising, or some step in the solution is. In this case, the twist is the fact that there are two possible sets of ages that add up to 14—and that this ambiguity, rather than preventing an answer, supplies it.

Insight Questions

Unlike regular logic puzzles, these questions are difficult to answer through step-by-step deduction. It takes a leap of insight. Either you hit on it or you don't. Those who don't have the *aha!* moment—and have it before the interviewer moves on—are out of luck.

Google sometimes uses this question:

> You've got a chessboard with 2 diagonally opposite corners chopped off—so it's 62 squares instead of the usual 64. You're given 31 dominoes, each exactly the size to occupy 2 adjacent squares. Arrange the dominoes to cover the chessboard.

The applicant will be wasting his time: the task is impossible. The challenge is to realize that and then prove it's impossible.

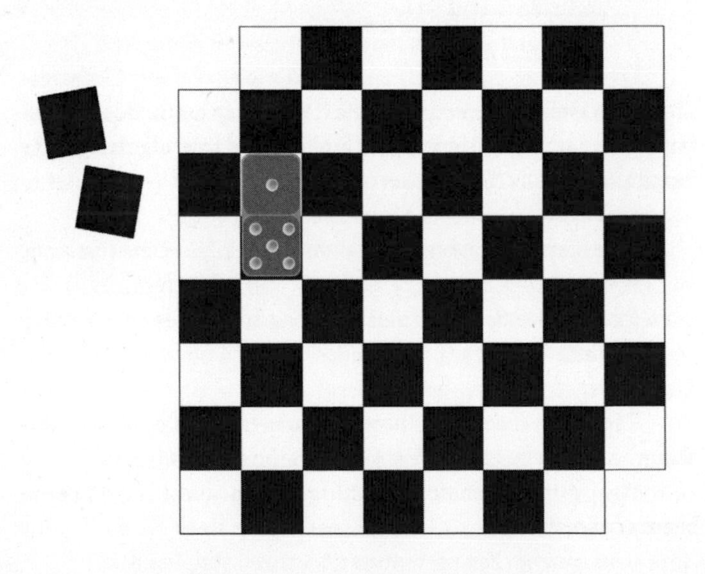

Insight questions are often amusing, for a solution is a "punch line" of sorts. But they put the candidate on the spot. There isn't much of a thought process to verbalize.

In this case, the solution depends on the observation that the two removed corner squares must be the same color. A domino, on the other hand, invariably covers one white and one black square. You can lay down thirty dominoes, covering sixty squares, and then you're going to be left with two uncovered squares of the same color. They can't be adjacent—at best, they'd be diagonally next to each other—and there's no way the remaining domino can cover them.

The best way to deal with these questions is to be aware of the commonly used ones. There's a finite number, most of them venerable. Martin Gardner mentioned this one in his *Scientific American* column in 1957.

Lateral Thinking Puzzles

Like a crazy uncle making the same joke at every Christmas dinner, some interviewers can't resist tossing these in. These questions hinge on verbal ambiguity and mainly test whether you've heard them before. Here's one:

> There are three women in swim suits. Two are sad, and one is happy.
> The sad women are smiling. The happy woman is crying. Explain.

The interviewers who use questions like this probably consider them to be fun. This demonstrates the relativity of fun.

The intended answer to the question above is, "They're beauty contestants."

Lateral thinking puzzles are short and don't seem to contain enough information to answer. That should tip you off.

Tests of Divergent Thinking

These are much like classic psychological tests for creativity. They are open-ended challenges that encourage brainstorming. There is no "right answer" to deduce. The goal is to toss out as many good ideas as you can, with extra credit for good original ideas.

How would you compare two search engines?

It often helps to begin by identifying a goal that may not be explicit in the question. A search engine is like a carnival fortune-teller, you might say. It is asked to read the user's mind from a few vague clues. The successful search engine (or fortune-teller) is good at convincing the user that it knows more than it does. The worst thing it can do is to respond in a way that the user knows is phony or feels to be "mistaken." Some typical good responses to this question include the following.

Measure how fast it is. How fast do search results appear?

Google yourself. Each of us is the world expert on our own self. By putting your own name into a search engine, you get a unique perspective on how relevant the links are.

Try a phrase made out of short, common words. Someone typing "to be or not to be" hopes to learn what play it's in, or perhaps to find out about Ernst Lubitsch's war-time comedy of the same name, which came out in 1942. The user doesn't want to be scolded that the search words are common and not searchable. Do the engines handle this better when the search string is enclosed in quotation marks?

Misspell a search term. "Exhibit pimp ride" should get links for the U.S. TV show (the host is Xzibit), not a museum display.

Check how it handles capitalization. A search engine can't require capitalization, for most users don't bother. It should make the most of any clues supplied, including capital letters. Check "googol" (the number), "Google" (the company), and "Gogol" (the author), each with and without the first letter capitalized. An uncapitalized "gogol" might lead the search engine to ask whether "googol" was intended.

Determine how hackable the ratings are. Every company wants its site to land high in the listings. There is an ever-changing technology of gaming the search results, engaged in a Red Queen's race with the search engines themselves. You could arrange a test with some site promoters, having them do everything they can to put a dummy site high in the listings. See whether the search engines fall for it.

Check susceptibility to Google bombing. In 2003, the blogger George Johnston started a campaign to make President George W. Bush the top search result for the search terms "miserable failure." Obviously, those words did not appear on Bush's whitehouse.gov page. Instead, Johnston urged bloggers to use the phrase and link it to the Bush page. This quickly achieved the intended result. Even typing "failure" and hitting "I'm Feeling Lucky" took Googlers directly to Dubya. The prank generated more hits for the Bush page than regular searches did. Search engines now try to resist Google bombing. Arrange a test and see how well the engines perform.

This interview question has its roots in a real test. In 1998, Larry Page and Sergey Brin pitched Google to the skeptical venture capitalist Ram Shriram. He insisted on a blind test. Shriram picked keywords and typed them into Google and the major search engines of the time. Google was the fastest, and Shriram ended up writing a check for $250,000.

Fermi Questions

A popular type of interview question asks for a quick, off-the-cuff estimate of an unknown quantity. The objective is not to get the "right" answer. For one thing, the interviewer won't know the right answer. The point is to demonstrate that you can work out a logical path to an answer. These questions are named for the physicist Enrico Fermi (1901–54), who used them in teaching.

"How many tennis balls can you fit in this room?"

Start by taking a quick look at the room and estimating its dimensions. A 10' × 10' × 10' office would be 1,000 cubic feet. Most job interviews take place in fairly modest offices, so the odds are you're looking at 1,000 to 2,000 cubic feet.

A regulation tennis ball is 2.575 to 2.700 inches in diameter. You aren't expected to know that, of course. Your offhand guesstimate is likely to be "about three inches across." Then you get about 4 balls to a foot, and something like 4 × 4 × 4 = 64 to a cubic foot, in a neat cubic lattice.

Multiply that (which you might round to "less than one hundred") by your estimate of the room's volume. Unless you're interviewing in the CEO's suite, the answer is going to be a hundred thousand or so.

Algorithm Questions

I'll use the term *algorithm questions* for a category that has become quite popular at many blue-chip companies as well as in Silicon Valley. They ask how you would perform a task, which ranges from the cosmic to the mundane. There is often a hint that efficiency counts—that you will be graded on how well your solution conserves time, effort, or money. Some algorithm questions are so vague, and so vaguely self-important, that the uninitiated might think they're expected to spout motivational slogans. Really, you're intended to take the question seriously and to come up with a detailed plan for performing the task. Here's an example:

> You have a wardrobe full of shirts, and it's very hard to find the one you want. How would you organize the shirts for easy retrieval?

"Get a wardrobe organizer" is not what the interviewer wants to hear. You're supposed to invent the wardrobe organizer.

With questions like this, it's usually best to start with a workable idea and keep improving it. Here, a promising first stab is to make a shirt rainbow. Arrange your shirts on a rack by color, in the order of the spectrum. You can take in the whole set of shirts at a glance and see exactly where every shirt is or ought to be. A kelly green shirt goes between the blue shirts and the yellow shirts, or more exactly between aqua and chartreuse.

On second thought, it's not that easy. You have to allow for nonspectral colors like white, tan, gray, and black. These could be appended to the end of the rainbow. You also have to allow for checked shirts, T-shirts with multicolor logos, and so forth.

Alternatively, you could arrange shirts by something other than color: sleeve length (long or short); style (pullover or fully buttoned); purpose (work, sport, casual, formal); fabric; designer.

Some of these distinctions are ambiguous. There's a thin line between work, sport, and casual wear, especially at Google. Fabrics can be blends, and not everyone pays attention to designers. The objective is to be able to glance at a shirt and immediately know exactly where it belongs.

One way to smash ambiguity with a mallet is to take the dry cleaner's approach and put a serial-number tag on every shirt. Arrange the shirts by serial number. The trouble with this is that unless these serial numbers are second nature to you, you'll have to do an awful lot of looking. How long does it take your dry cleaner to find your shirt? If the serial numbers are assigned without rhyme or reason, the shirts are essentially in random order. That's no improvement over having no "system" at all.

Here's a practical solution: Divide your shirts into as many nonambiguous categories as you can. Allow a wardrobe-organizer bin (or a section of rack) for each category. The optimal categories will depend on your wardrobe. Here is an example:

Solid colors: purple, blue, green, yellow, orange, red/pink, white, gray, black, tan/brown (10 bins).
Stripes (1 bin).
Checked (1 bin).
T-shirts and jumpers with printed slogans, by first letter of slogan: A–L and M–Z (2 bins).
T-shirts with pictures, no text (1 bin).

The disadvantage of the bin approach is that there will sometimes be more than one shirt in a bin. This might not be an issue for a salaryman with twenty light blue oxford-cloth button-down shirts, all interchangeable. Just grab any shirt out of the bin, and when it comes back from the cleaners, replace it in the bin, anywhere. However, interviewers expect you to be able to deal with the more common case of needing to find a specific shirt. You may have just one Manchester United T-shirt, in the bin labeled "M–Z (T-shirts and jumpers with printed slogans)." You need that specific shirt, and maybe there are lots of other shirts in there. How do you efficiently arrange the shirts within the bin?

The obvious answer is to arrange the shirts within each bin in a suitable linear order. This could be alphabetical order, spectrum order, or anything else. This is an efficient, scalable solution (in theory, anyway).

A smarter, more realistic answer might be to invoke the 80-20 rule for shirts. You probably wear 20 percent of your shirts about 80 percent of the time. Most garments are rarely worn (but are "too good to throw away").

To take advantage of this, design your bin categories so that there will be only a few frequently worn shirts in each bin. You can always do this: Should one category have too many shirts, split it into smaller, more restrictive categories (like "short-sleeve blue shirts" and "long-sleeve blue shirts"). The most popular shirt goes on top of each bin's stack. It's easy to find and replace; you don't have to rummage underneath it except when looking for a less popular shirt. The other shirts within each bin should be arranged in descending order of popularity. This way, it's easiest to access the shirts you're most likely to want.

Coders asked such questions are expected to see parallels to software design. The sorting of shirts by color or pattern is a *hash function,* and the linear arrangement of shirts within a bin allows a *binary search.* For everyone else, algorithm questions are informal tests of organizational ability. In answering, it's vital to understand what you're economizing (in this case, the time or effort required to find a shirt).

QUESTIONS

Identifying the Genres

Here's an assortment of tricky interview questions exemplifying the popular genres. First try to decide what kind of question each is, and then answer it. (Answers start on p. 198.)

? How much would you charge to wash all the windows in Seattle?

? A man pushed his car to a hotel and lost his fortune. What happened?

? You get on a ski lift at the bottom of the mountain and take it all the way up to the top. What fraction of the lift's chairs do you pass?

? Explain what a database is to your eight-year-old nephew, using three sentences.

? Look at this sequence:

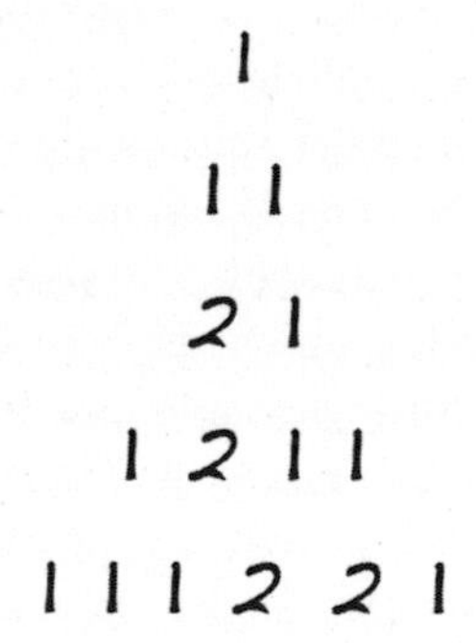

What's the next line?

? You have twenty-five horses. How many races do you need to find the fastest three horses? You don't have a timer, and you can run only five horses per race.

Seven

Whiteboarding

The Art of the Visual Solution

Whiteboarding—mordantly echoing *waterboarding*—is the practice of having job candidates write or diagram their thoughts as they answer a difficult question. Used extensively in today's interviews, it's a form of psychoanalysis, forcing candidates to lay bare their most private thoughts to a not always sympathetic audience. Whiteboarding is mandatory with very technical questions, but it can also be useful with the genres discussed in the previous chapter. "Even if it's not a coding question, draw out your logic on the whiteboard," advises Google's Todd Carlisle. It's helpful with questions having an obvious visual element and often with those requiring elaborate deductions.

Drawing or writing gives you something to do with your hands while hoping inspiration strikes. You might start by jotting down the main features of the problem, just to make sure you understand them. The whiteboard is also a memory aid, a place for recording conclusions or intermediate values. Best of all, drawing a diagram is often genuinely helpful in solving the problem. Here's an example, asked at Google.

Break a stick at random into three pieces. What's the probability the pieces can be put together to form a triangle?

The first thing you need to understand is how the three pieces can *fail* to form a triangle. You'd probably begin by drawing something like this:

When you break a stick into three roughly equal pieces, they will *always* form a triangle. It won't necessarily be a neat, regular triangle, but it will have three sides and three angles.

But three stick segments can fail to make a triangle. Here are two examples, on the following page. In each case, one side is longer than the other two sides put together. There is no way the two short sides can bridge the long side.

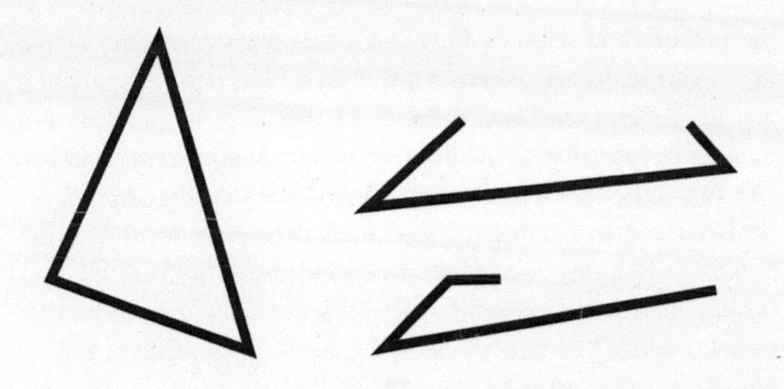

Call the original stick's length 1 unit. If the largest broken piece is more than 0.5 units long, the other two segments will together be less than 0.5, and they won't form a triangle. Otherwise, they will. It's that simple.

The question therefore becomes, what's the chance that the largest stick fragment will be no more than half the length of the original stick? The answer depends on what you mean by "break a stick at random." It's appropriate to ask the interviewer for clarification.

Were a magician to pull me out of an audience and tell me to break a stick "into three random pieces," I would grab the stick in two fists and bend it until the wood snapped. I wouldn't try to break it into equal halves, nor would I try to make an extremely lopsided division. I would then select the larger of the two pieces (provided one was conspicuously larger) and break it the same way.

This procedure isn't especially random. The physics of wood fibers and the size of the human fist bias the breaking. In a truly random breaking, *any* division of the stick ought to be possible. It could conceivably produce two tiny stubs and one big stick virtually as long as the original. This would not form a triangle.

The interviewer will grant you license to forget the practicalities and assume a mathematically random breaking. (Actually, the interviewer *has* to say that because he hasn't a clue how to address the practicalities, either.) That yields this reinterpretation of the question. Pretend the stick is a meter stick, with one end marked 0 and the other 1, and with labeled tick marks throughout. Pick a number between 0 and 1, using any random device or function you like. Break the stick at that point. Then pick another random number between 0 and 1. Break at that point. (The point for the second break may fall on either of the two sticks resulting from the first break. That's why it's handy to use a stick marked like a ruler.) This would produce three truly random segments. Each segment may be anywhere between 0 and 1 in length.

This interpretation allows a simple diagram. Chart the position of the first break on the *x* axis and the second break on the *y* axis. This makes a square. Every point in the square represents a possible way of randomly breaking a stick in three. All points are equally likely, so areas of the chart correspond to probabilities.

Here's the diagram (for clarity, a little fancier than what you'd draw during an interview).

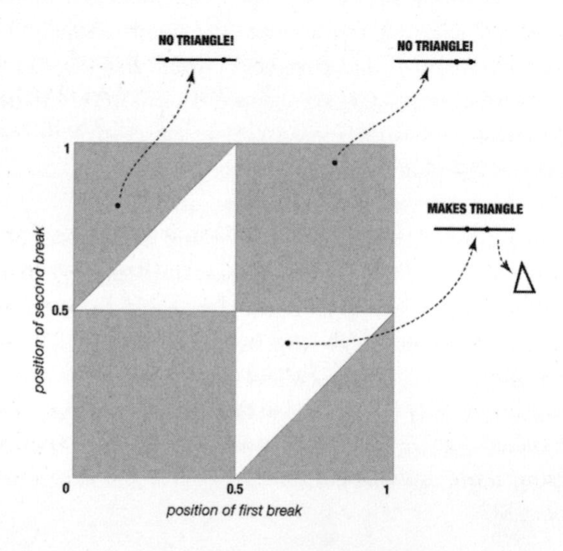

Divide the square in half both ways to make four smaller squares. The cases where both breaks are on the same side of the midpoint (0.5) are represented by the southwest and northeast quadrants. I've shaded these square regions to indicate that the resulting pieces of the stick won't form a triangle.

The other two quadrants are cases where the breaks fall on opposite sides of the midpoint. These can be divided in two diagonally. In the shaded triangular regions, $x - y$ or $y - x$ is greater than 0.5. That means the middle segment is too big to allow a triangle. In the bow-tie-shaped white region, the middle segment is less than 0.5, and the pieces can make a triangle. It's easy to see that the white regions are one-fourth the chart's total area. Therefore, the chance of breaking a stick into three random pieces that form a triangle is one in four.

Pictures and Mappings (and Pizza)

In the puzzle above, I used the whiteboard in two ways: to draw a literal *picture* of the broken sticks and to make a conceptual *mapping* of the chances. Interviewers are especially impressed with those who can use mappings to solve a problem visually. The whiteboard can be useful even when you don't need to draw a picture of anything. Here's an example.

> You and a friend are sharing a pizza. It takes you X seconds to eat a unit of pizza, and it takes your friend Y seconds. The rules of pizza etiquette say that you can eat only one slice at a time. You can't reach for a new slice until you've finished the one you're eating. Should you and your friend reach for the last slice at the exact same instant, he gets it (the "tie-breaker rule"). The pizza must be cut into equal slices. How many slices should there be to give you the most pizza possible?

Your first impulse might be to draw a pizza cut into slices. Well, okay, no harm in that. But it probably won't get you far. This problem is more about time than pizza.

As everyone knows, the faster pizza eater has an advantage. He gets seconds before anyone else has had firsts. This is true whether the pizza is sliced into eight pieces, Domino's-style, or another number, whether the pizza is round or rectangular, thick crust or thin.

Examine some extreme cases. Suppose you didn't slice the pizza at all, leaving it whole. You and your friend both reach for it, and the tiebreaker rule kicks in. The first "slice" is also the last, and he gets it—the whole pizza!

You definitely want to slice the pizza. The second-simplest case is to slice the pizza into halves. Each of you gets one slice, obviously, and that's a fair fifty-fifty split. That's an important thing to know. You never need settle for less than 50 percent of the pizza.

Another extreme: cut the pizza into an infinite number of infinitesimal slices. In the limit of many, many slices, it scarcely matters who gets the last microscopic sliver. The net effect is to split the pizza by eating speed. Should you and your friend eat at the same rate, you'll each consume half the pizza. Should you eat twice as fast, you'll get twice as much pizza.

That suggests a strategy:

- If your friend eats faster than you, slice the pizza in two. This guarantees you half the pizza.
- If you eat faster, cut the pizza into an infinite number of slices. This guarantees you a split by eating speed (so you get more than half).

It's "heads I win, tails you lose."

You should never be quick to settle for the first workable answer. Can you do better?

One flaw of the above strategy is this business about cutting the pizza into an infinite number of slices. Slicing would consume an infinite amount of time—after which the sun will be a burned-out cinder and the pizza will be stone cold. Even if you settle for "a lot" of slices, as an approximation to infinity, you waste "a lot" of time.

There is a much better strategy. Draw the slice-grabbing timeline. You and your friend start out with one slice each. Then the faster eater finishes his slice first and grabs a second. Stop the clock there: at that instant, the faster eater has two slices and the slower eater just one.

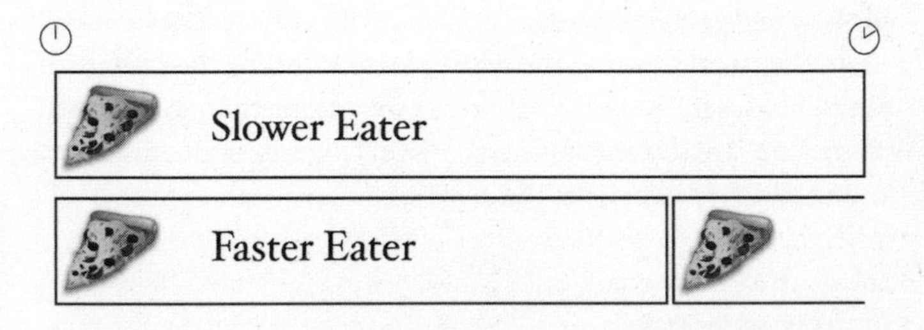

What if these three slices were all there were? Then the faster eater would end up with two-thirds of the pizza. Notice that the faster eater doesn't have to be *much* faster. He could be a split second faster. This is much better than the infinite-slices scheme, where a slightly faster eater gets only slightly more than half.

Suppose the faster eater is more than twice as fast as the slow guy. He can polish off two slices and begin on a third while the slow guy is still on his first. If it's a four-slice pizza, the faster eater would get three-quarters of the whole. For the faster eater, it's all about getting seconds (thirds, fourths...) while the slower eater is still on his first slice.

The best strategy, then, is to figure out how many slices you can grab while your friend is still on his first piece. According to the question, you take X seconds to eat a unit of pizza, and your friend takes Y seconds. The bigger Y is, the more time you'll have to stuff your face. Specifically, you'll be able to grab Y/X slices (rounded up, if necessary, to an integer) while your friend is on his first. Divide the pizza into that many slices *plus one.* This guarantees you'll get all but one of the slices.

I'll make it easier for you. Realistically, X and Y are unlikely to be too far apart. One person may take 30 percent longer to finish a slice, but he's not likely to eat three times faster, or be 1/10 as fast. There are two plausible situations. When you're the slower eater, cut the pizza in two and get half. When you're faster, cut it in three and get two-thirds.

The secret to using the whiteboard is finding the right mappings. (That's the secret to solving any problem, though the "mappings" are usually mental ones.) Don't be shy about using the whiteboard's best feature: the eraser.

QUESTIONS
Visual Solutions

Here are some more interview questions that can be solved by drawing a picture or a diagram. (Answers start on p. 205.)

? Imagine you have a rotating disk, like a CD. You are given two colors of paint, black and white. A sensor fixed to a point near the disk's margin can detect the color of paint underneath it and produce an output. How do you paint the disk so that you can tell the direction it's spinning just by looking at the sensor's readings?

? How many lines can be drawn in a plane such that they are equidistant from three noncollinear points?

? Add any standard arithmetic signs to this equation to make it true.

$$3 \quad 1 \quad 3 \quad 6 = 8$$

? There's a bar where all the customers are antisocial. It has 25 seats in a line. Whenever a customer comes in, he invariably sits as far from the other customers as possible. Nobody will sit directly next to anyone else—should a customer come in and find that there are no such seats, he walks right out. The bartender naturally wants to seat as many customers as possible. If he is allowed to tell the first customer where to sit, where should that be?

? How many different ways can you paint a cube with three colors of paint?

Eight

Dr. Fermi and the Extraterrestrials

How to Estimate Just About Anything in Sixty Seconds or Less

"I'm going to ask you a few questions that may sound strange," announced the voice on the phone. His name was Oliver, and he was interviewing Alyson Shontell, a Syracuse University advertising student, for an assistant product manager position. "These questions are meant to test your analytical thinking. I want you to estimate how much money you think Google makes daily from Gmail ads."

"Um, you mean a hard number?" Alyson asked. "Maybe... seventy thousand dollars?"

Oliver got a good laugh out of that one. *Of course* he wanted a hard number, and $70,000 wasn't it.

"Wait, can you just totally ignore that response?" Alyson asked. "Scratch it out of your notes and pretend I never said that?"

"You don't have to give me an exact number, just tell me how you would figure out the answer."

"Okay," Alyson forged on, "Google places four ads per e-mail opened in Gmail.... Say each Gmail user opens seven new e-mails a day. They would see twenty-eight ads. If they click on one-fourth of those ads, then only seven ads are clicked. If all advertisers are charged five cents per clicked ad, then the amount of revenue would be whatever five cents times seven ads times the number of Gmail users is.... Does that make any sense at all?"

"Kind of," Oliver said uncertainly. "You lost me at the 'only clicking on one-fourth of the ads' comment." Then: "Let's move on."

Alyson's second interview, also by phone, was conducted by Anna. "Name a piece of technology you've read about recently," Anna began.

That was easy! "Okay, today I was reading about Nike and Apple working together to make a shoe with a chip in it that helps you run in time with your music."

"Now tell me your own creative execution for an ad for that product."

Alyson sketched a perfect word picture of an ad. It had a runner, an iPod, and a dramatic crossing of the finish line. Anna was chuckling faintly—because it was stupid, or because it was brilliant?

"Now I'm going to ask you a maths problem," Anna said. "Say an advertiser makes ten cents every time someone clicks on their ad. Only twenty percent of people who visit the site click on their ad. How many people need to visit the site for the advertiser to make twenty dollars?"

"Um... well, okay. So, twenty out of one hundred people click on the ad. Every ten clicks make one dollar... and you need twenty of them..." This wasn't going well. Alyson began guessing, blindly flailing about.

Finally, Anna explained the correct answer. It takes five site visitors to generate one click worth ten cents, she said. Twenty dollars is two hundred times that, so it would take five times two hundred, or a thousand visitors.

Alyson felt like a complete moron. Anna felt like it was time to ask another maths problem.

"Estimate the number of students who are college seniors

[in their final year at university], attend four-year schools, and graduate with a job in the United States every year," she demanded.

"There are about 300 million people in the nation," Alyson said. "Let's say 10 million of those are college students at four-year schools. Only one-fourth of those 10 million are seniors, so that would be roughly 2 to 3 million. If half of those students graduate with jobs, you're looking at about 1.5 million kids."

"Would you say that number seems high, low, or just about right?"

"I would say it sounds low, but maybe that's because I'm going through the job-search process and I'm wishing the number was higher."

Not even a laugh.

"That's all," Anna said. "Good luck with your job search."

Lunch at Los Alamos National Laboratory

Blame the extraterrestrials for this style of interviewing. In 1950 the subject of flying saucers came up one day around lunchtime at Los Alamos.

"Edward, what do you think?" the physicist Enrico Fermi asked his tablemate Edward Teller. Was it possible that extraterrestrials were visiting the earth in spaceships? Teller judged it highly unlikely. Fermi was not so sure. He spent much of his lunch calculating how many extraterrestrial civilizations there were in the universe and how close the nearest one would be.

This was the classic "Fermi question." Back at the University of Chicago, Fermi tormented his students with only somewhat easier questions. His most famous classroom riddle was, "How many piano tuners are there in Chicago?" Fermi staunchly believed that anyone with a PhD in physics should be able to estimate just about anything. Somewhere along the line the "PhD in physics" part got dropped.

Today's employers have gotten the idea that everyone, including humanities graduates, should be able to estimate odd quantities on a job interview. (No one is expected to estimate weird things after they're hired.) These questions are today's riddle of the sphinx. They often determine who gets to pass from phone interview to the corporate campus. Some are loosely related to the company's line of business:

How many petrol stations are there in the United States? (asked at General Motors)

But more often, there's no discernible connection:

How many rubbish collectors are there in California? (Apple)
Estimate the number of taxis in New York City. (KPMG)
How many golf balls would fit in a stadium? (JP Morgan Chase)
Estimate all the costs of producing a bottle of Gatorade [a U.S. sports drink]. (Johnson and Johnson)
How many vacuum cleaners are made a year? (Google)

An advantage of Fermi questions—for employers—is that it's easy to invent new ones. Thus the candidate can be presented with a fresh question that's never been in a book or on the web. Intel is especially clever: its interviewers sometimes ask engineers to estimate the number of lines of C or C++ code they've written. This has at least three agendas. First, it's a Fermi question. Second, it asks how much coding experience the applicant has. Finally, since the answer involves estimates of hours worked, it's a sneaky way of asking whether the candidate is willing to work nights and weekends.

A drawback of this free invention is that interviewers may

not know how difficult a newly minted Fermi question is, much less know the correct answer. Sometime after her phone interview, Alyson Shontell ran into Oliver at a conference. He confessed that he didn't know how much Google makes from Gmail ads.

"The interview was going swimmingly until I met up with one interviewer who was apparently antimilitary," recalled one unsuccessful candidate, a six-year vet of army intelligence. "Using the Google 'Do no evil' mantra as a pretense, he asked me how many people I'd killed when I served. When I explained to him that I was [military intelligence], he then asked if I could estimate how many people were killed because of the intelligence I'd gathered. The implication was I was either an evil, efficient killer or an incompetent one."

The Los Alamos crowd might have had trouble with that ethical equation. Fermi, at any rate, believed in the interconnectedness of knowledge. It was his observation that someone with a critical mass of facts and figures could use that to estimate very exotic things, like the number of light-years to the nearest alien civilization. These questions are therefore about charting a reasonably direct path from *facts you know* to the weird statistic demanded. That means you have to know some facts and be able to draw connections. You can round the maths, but you shouldn't pull numbers completely out of the air.

That was Alyson's problem in the Gmail question. Instead of starting from something she knew, she tried to guess a flurry of unknown and proprietary statistics. (Google plays its cards close to the chest on the financials of specific products.) A better approach would have been to start with two basic balance-sheet facts. They are (a) that Google makes most of its money from advertising, and (b) its annual revenue is about $25 billion.

You may be wondering, "How in the world would I know what Google's revenue is?" The answer is, you look it up before the interview. It's one of the few financial statistics you should know when interviewing at a large company.

How much of that $25 billion is from Gmail ads? You have every right to draw a blank there. It's something that the financial press speculates on. Ideally, you would have read enough about Google's business to know that Gmail ads are a fairly insignificant part of the whole. When Gmail was rolled out in 2004, there was hopeful talk of Gmail ads eventually being as important as search ads. It hasn't worked out that way.

It's unlikely that an interviewer would fault you for saying, "Let's guess that Gmail is one percent of the total revenue. If it's more or less than that, I can easily adjust it." This also makes the maths easy. The Gmail revenue would come to $250 million a year. Divide that by 365, and you've got something under $1 million a day. That illustrates the cardinal rule of Fermi questions: round corners on the arithmetic, not on the logic.

Alyson started off well with the university question. She knew the U.S. population is about 300 million. Her reasoning went off the tracks when she said, "Let's say 10 million of those are college students at four-year schools." She threw out an accurate statistic and pulled a new fantasy number out of the air.

A better reply would run like this. The U.S. population is about 300 million, and the life expectancy is about 70 years. Those two numbers imply that, each year, something like 300 million / 70 pass a given age milestone—such as their 21st birthday.

You can tweak that before you do the division. The number of Americans turning 21 will actually be bigger because more people make 21 than 70. How much bigger? Most likely, neither you nor the interviewer knows. This provides cover to simplify the maths. Answering these questions is like parkour. You want to segue suavely from one conveniently round figure to another without a stumble. Instead of 300 million / 70, let's call it 300 million / 50. That would come to 6 million Americans turning 21 each year.

Not all are college seniors. Many Americans don't go to university; others do but take two-year degrees or leave before their final year. Against this should be balanced the fact that some who are older than traditional college age matriculate and become seniors. Use that to trim the 6 million figure to 3 million college seniors.

The next part of the question is, how many seniors are in four-year courses in university? As is obvious, if you know the U.S. university system, all college seniors are in four-year universities. The final part of Anna's question was, how many seniors graduate *with a job?* This could refer to those with a solid job offer even before graduating. A broader interpretation might be getting a job in a reasonable time (whatever that means in today's job market). You should ask the interviewer for details.

Typical responses:

"A job waiting on graduation day": 25 percent of seniors, or 750,000

"Getting a job in a reasonable time": 50 percent of seniors, or 1.5 million

You'll notice that 1.5 million is the exact number Alyson gave. Meet the flip side of "you don't have to get the right answer." You can fail even when you do get a "right answer." It's the journey, not the destination.

An Interview Cheat Sheet

Interviewers don't expect candidates to have lots of obscure statistics on the tips of their tongues. The only things it's embarrassing not to know are important figures within one's field of expertise, some very basic demographic facts, and a few statistics about the company. Here's a manageable cheat sheet of round figures (replace the Google/United States data with that of the company/country where you're interviewing):

Population of the world: 7 billion
Gross domestic product of the world: $60 trillion
Population of the United States: 300 million
Gross domestic product of the United States: $14 trillion
Federal minimum wage: $7 (actually $7.25)
Population of the San Francisco metro area, including Silicon Valley: 8 million
Google's stock market value: $100 billion
Google's annual revenue: $25 billion
Google's annual profit: $10 billion
Price of a share of Google stock: $600
Number of spheres that fit in a big volume, with random packing: 1.2 times what you would compute assuming a cubic lattice (for more on this, see p. 220)

QUESTIONS

Impromptu Estimates

For practice, here are several more Fermi questions. (Answers begin on p. 218.)

? How many ridges are there on the rim of a U.S. quarter?

? How many bottles of shampoo are produced in the world in a year?

? How much toilet paper would it take to cover Greater London?

? What is 2^{64}?

? How many golf balls will fit in a Mini?

Nine

The Unbreakable Egg

Questions That Ask "How Would You . . . ?"

Headmaster Douglas Appleton of Carr Mill Primary School, Lancashire, created a media sensation in 1970. All he did was to demonstrate a simple, counterintuitive fact: a raw egg dropped onto grass will usually *not* break—regardless of the height of the drop. Appleton's students tossed eggs out of the school's second-story window. The eggs that fell on grass didn't break. An obliging fireman then climbed to the top of a 70-foot ladder and dropped ten eggs onto grass. Seven of the ten were unharmed. An RAF officer followed with a similar experiment, dropping eighteen eggs from a helicopter 150 feet above the ground. Fifteen (83 percent) survived. A newspaper, the *Daily Express,* rented a small plane to dive-bomb sixty eggs onto a lawn while traveling at 150 miles per hour. About 60 percent of them made it intact.

I mention this to prove that one of Google's interview questions is not completely preposterous.

> You work in a 100-story building and are given two identical eggs. You have to determine the highest floor from which an egg can be dropped without breaking. You are allowed to

break both eggs in the process. How many drops would it take you to do it?

Lest there be any confusion, the building and eggs are strictly imaginary. This is an algorithm question. It's a test of your ability to craft a smart, practical way of doing something. That's important in engineering, in management, and in everything else.

Every cook knows that a raw egg dropped from counter height onto tile is history. But if Google's 100-story building is surrounded with something softer than concrete and harder than grass, the answer is not obvious. To answer in the spirit intended, you have to assume that it is possible that the maximum egg-safe floor could be any floor at all, from 1 to 100. In fact, you should allow for the possibility that no floor is egg-safe (seconding the wisdom of cooks).

You are permitted to ignore the strong element of chance in egg defenestration (demonstrated in the 1970 experiments). Pretend that the outcome of dropping an egg from a given floor will always be the same. Either it breaks or it doesn't.

The interviewer is not expecting you to name the egg-safe floor. There are no eggs and there is no 100-story building: it's all a fictional situation, okay? You are asked only to devise an efficient method for determining the floor, while explaining your thought process. The one question you *are* expected to answer with an actual number is, how many drops would you need? Scoring is like golf: the fewer, the better.

Bits and Eggs

Dropping an egg is a simple experiment that yields one bit of information. To get the most bang for the bit, you'd start in the middle of the building. That would be floor 50 or 51, as there's no exact "middle" floor in a building with an even number of floors. Try dropping an egg from the 50th floor. Let's say it breaks. This would mean that the desired egg-safe floor must be below the 50th floor. Split the difference again by dropping the next egg from the 25th floor. Oops! It broke again. Now you're out of eggs. You can now conclude that the highest egg-safe floor is below the 25th. You don't know which floor, and that means that the method has failed.

It's possible to keep reusing an egg so long as it doesn't break. Start on the bottom floor and drop the first egg. If it survives, go to the 2nd floor and try that. Then the 3rd, the 4th, the 5th, and so on until the egg breaks. That will tell the highest floor the egg can be dropped from without breaking. You will have determined it with just one egg.

Call this the "slow algorithm." It's a miser of eggs and a spendthrift of egg drops. Every single floor might have to be tested, but it gets the job done.

The challenge is to create a solution that makes good use of both eggs. Suppose that Google's optimal algorithm for the egg-drop experiment was written in a book somewhere. You don't even have to suppose—it *is* written in a book, namely this one. Turn to page 116 (uh, you're already there) and take a peek at how the algorithm begins. It reads something like this:

1. Go to floor *N* and drop the first egg.

How do I know the algorithm begins this way? Well, I'm not going out on much of a limb. An algorithm is a list of idiotproof instructions, starting with instruction 1. It tells you to drop an egg, naturally, because that's the modus operandi here. There's nothing else to do but drop eggs. The only interesting part (floor *N*) is currently concealed under an algebraic veil. The *real* algorithm gives a specific floor, like 43, in place of the *N*.

Further deduction: because the experiment in instruction 1

has two outcomes, it has to have follow-up instructions for both eventualities. Call them 2a (what to do if the egg breaks) and 2b (what to do if it survives).

Once you break the first egg, you will have to play it safe with the second. You can't take the risk of skipping any floor, lest you break the second egg and not be able to deduce the correct floor.

Instruction 2a of Google's algorithm has to say this, in so many words:

> 2a. (To be used if the egg breaks in 1.) Go down to floor 1. Adopt the "slow algorithm" with the remaining egg. Test it from every floor, working your way up the building until the egg breaks. The maximum egg-safe floor is the one below that.

For instance, imagine the first drop is from the 50th floor, and the egg breaks. You can't risk trying the second egg from floor 25 because it might break, too. Instead, you have to try floors 1, 2, 3 . . . potentially all the way up to 49. And since we started with the 50th, that could mean making fifty drops in all.

It doesn't take much coder's intuition to see that a search needing fifty tests to find one thing out of a hundred is *not* optimal. It's lousy.

It's better to count on making the first drop from a lower floor. If we start from floor 10, and the egg breaks, we might need as many as ten drops. This is the key *aha!* moment of the puzzle. In general, when the egg breaks on a first drop from floor N, it will take as many as N drops total to identify the right floor.

This strongly suggests that the first drop should be from a floor much lower than 50. Try $N = 10$. This choice has the appealing feature that we use the first egg to deduce the tens digit of the egg-safe floor, and the second egg to find the ones digit. For example, test the egg from floors 10, 20, 30, 40, and 50. Say it breaks

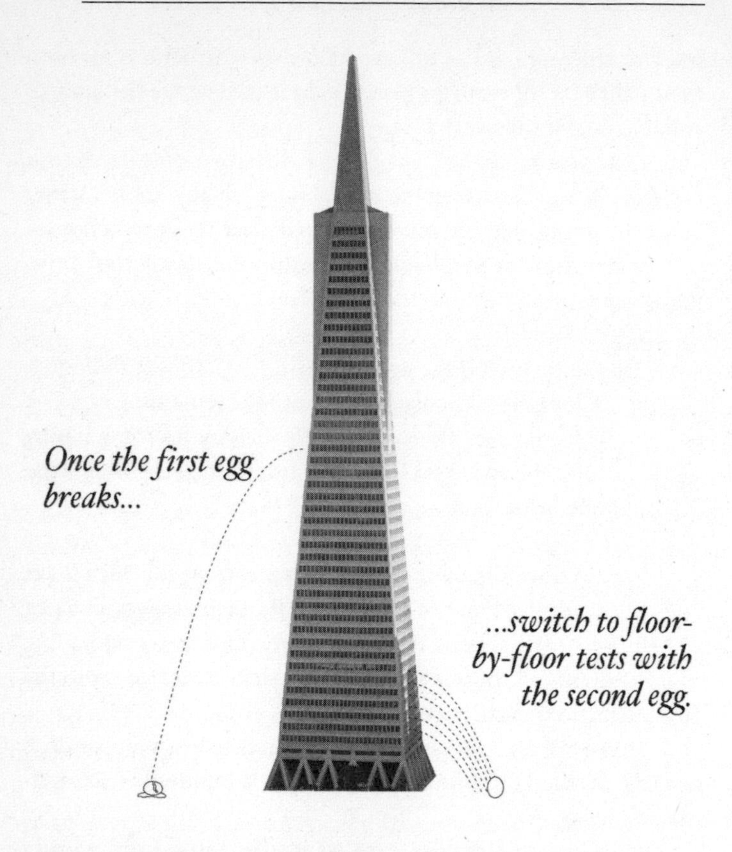

when dropped from floor 60. This tells us the maximum egg-safe floor is 50-something. Move down to floor 51 and work upward floor-by-floor with the other egg. If the second egg breaks at floor 58, that means the egg-safe floor is 57.

How efficient is this procedure? The worst-case scenario would be to try 10, 20, 30, all the way up to 100, where the egg finally breaks. Then you'd backtrack to 91 and work up. You could end up needing nineteen drops in all to determine that 99 is the correct floor.

This isn't a bad approach. But it's not the best.

Crash Test

Remember, the question asks, "How many drops would it take you to do it?" That's a strong hint that you're being graded on how few drops you need. (More exactly, you're to minimize the number of drops required in a worst-case scenario. Sometimes you'll get lucky and arrive at the answer in just a few drops.)

Because the first egg's role is as crash-test dummy, you want to put it in high-risk situations; that's how you learn as much as possible as quickly as possible. The second egg is a backup. Once it's the sole remaining egg, you have to play safe with it.

It's the crash-test-dummy egg that's crucial to achieving a good solution. It's the egg that can eliminate many floors in a single drop. The question is, how many? A bit of mental gymnastics is required to answer that. It throws many smart people. I'll begin with an analogy. You're a professional golfer on the eighteenth hole, vying for a big prize. To win, you've got to make the hole within three strokes. That necessity dictates which clubs you choose and whether you risk a bunker rather than playing it safe. It will require you to aim for the hole on the third stroke (rather than being satisfied with making the green). The three-stroke limit constrains your strategy throughout.

Google's perfect algorithm also has a limit: a maximum number of drops needed to determine the correct floor. Call this number *D*. The *D*-drop limit constrains your strategy.

For the sake of concreteness, imagine the limit is ten drops. Then you might as well drop the first egg from the 10th floor. See why? You want to choose a floor as high as possible, to rule out as many floors as possible. The 10th floor is the highest option, for this reason: should the first egg break, you could end up needing all ten permitted drops to determine the correct floor.

(The above paragraph is the hardest thing to understand, I promise.)

Everything else follows from this insight. After the initial drop, you've got nine drops left. Assuming the egg survives, you will again want to move up as many floors as possible for the second drop. You might think you'd move up another 10 floors. Not quite. Since you've only got nine drops left, you can move up 9 floors at most. That's because the egg could break on the second drop, forcing you to use a floor-by-floor search thereafter. You might have to test every floor between 10 and the one you just tried, 19, using up all your allotted drops. Had you gone up even a floor higher, you could get caught short and not be able to single out the correct floor.

Or say the egg survives the first two drops. That leaves eight drops. You would go up 8 floors for the next drop.

The floors you test, assuming a series of unbroken eggs, form a simple series.

10

$10 + 9 = 19$

$10 + 9 + 8 = 27$

$10 + 9 + 8 + 7 = 34$

etc.

Wait a minute. The highest floor you can possibly reach is $10 + 9 + 8 + 7 + 6 + 5 + 4 + 3 + 2 + 1$. That comes to 55. This scheme would work perfectly for a 55-story building. But the question says it's a 100-story building.

That's easily fixed. Remember, I just picked 10 out of the air. Replace 10 with *D*, the required number of drops in the best algorithm. The highest floor an optimal method can reach is

$$D + (D - 1) + (D - 2) + (D - 3) + \cdots + 3 + 2 + 1$$

This must equal or exceed 100.

From here on it's just algebra. The sum above is D plus every whole number smaller than it. This is a *triangular number.* Picture a rack of billiard balls. It's 5 + 4 + 3 + 2 + 1 balls. You may recall from school that the total can be computed by multiplying 5 by 5 + 1 and dividing by 2. So you've got 5 × 6/2 = 15, and that's the number of balls in a rack.

In this case, the sum of D and every smaller whole number is equal to D times $(D + 1)$ divided by 2. So:

$$D \times (D + 1)/2 \geq 100$$

Multiply both sides by 2, and you get

$$D^2 + D \geq 200$$

Focus on D^2 and ignore the much smaller D. This equation says that D^2 is at least as big as 200 or so. The square root of 200 is just over 14. Try that for D.

$$14^2 + 14 = 196 + 14 = 210 \geq 200$$

Bingo. It fits. Just to be sure, try 13.

$$13^2 + 13 = 169 + 13 = 182$$

Nope, that's not greater than or equal to 200. Fourteen it is. You drop the first egg from floor 14, and you're guaranteed to find the answer in fourteen drops or less.

The recap: First, drop the egg from floor 14. If it breaks, go down to floor 1 and work your way upward floor-by-floor. This will get the answer in no more than fourteen drops total.

Should the first drop *not* break the egg, go up to floor

27 (14 − 1 floors above floor 14) and try again. Should it break this time, you'll have to go down to 15 and work up. This will also yield the answer in fourteen drops total or less.

Given a string of unbroken eggs, you'd test floors 39, 50, 60, 69, 77, 84, 90, 95, 99, and then 100 (actually, were the building higher, the next floor would be 102). This means it would take twelve drops, and no broken eggs, to deduce that the egg can survive every floor of the building. Should the egg break at any point in the process, you'd be shunted into the slow algorithm and might need all fourteen permitted drops.

How to Spot an Algorithm Question

Technology companies started asking algorithm questions as a way of testing engineer applicants' ability to apply what they'd learned. Instead of posing another coding exercise, they invented a more or less amusing story problem. This type of question has since spread to interviews for nontechnical positions. A secretary or manager is always figuring out how best to do something while minimizing time, money, or effort.

Hints that you're dealing with an algorithm question are

- a silly task that usually has nothing to do with the work for which you're applying;
- a weird constraint on that task (like using just two eggs);
- a goal, explicit or not, of economizing something (in this case, the number of egg drops); and
- an algebraic *N*, a big round number like a hundred or a trillion, or an indefinite quantity (a hint that the interviewer wants a scalable solution).

Efficient is a word that everyone uses, and it can mean many different things. If the question doesn't make it clear what the constraints are, it's appropriate to ask the interviewer. Another piece of advice that's often useful with these questions: start simple. When the question mentions a large number (like 100 floors), consider how you'd tackle the simplest cases (a 1-story building, a 2-story duplex, etc.). The answer is usually obvious. Then work your way up to larger values. A pattern will often emerge that you can extend.

There's usually more than one approach to an algorithm question. Your first workable idea may not be the best one, so you should be willing to explore alternate approaches (unless the interviewer is delighted with your first idea and moves on to the next question).

Speaking of alternate approaches, a reader on the Classic Puzzles blog supplied this answer to the egg question:

1. Drop egg from second floor. Watch it break. Curse under your breath.
2. Go down to first floor, drop second egg. Soon realize that even a 1 story drop is too much for an egg.
3. Degrade your interviewer and dare him to get an egg to not break more than 1 time in a row from any freaking window in that whole stupid building.
4. Get promoted.

QUESTIONS

How-To Puzzles

Here is a selection of algorithm questions, most requiring no special knowledge. The last two questions here—"point A to point B" and "closest pair of stars"—are asked of software engineers at Google and are more technical than the others in this book. I include them because they sound like Zen koans and occasionally inspire clever answers other than the intended one. (Answers begin on p. 222.)

- It's raining and you have to get to your car at the far end of the car park. Are you better off running or not, if the goal is to minimize how wet you get? What if you have an umbrella?

- You have a glass jar of marbles and can determine the number of marbles in the jar at any time. You and your friend play this game: In each turn, a player draws 1 or 2 marbles from the jar. The player who draws the last marble wins the game. What's the best strategy? Can you predict who will win?

- You've got a fleet of fifty lorries, each with a full tank of petrol and a range of 100 miles. How far can you deliver a payload? What if you've got *N* lorries?

- Simulate a seven-sided die with a five-sided die. (*Die* is the singular of *dice,* in case you didn't know.) How would you produce a random number in the range of 1 to 7 using a five-sided die?

- You have an empty room and a group of people waiting outside the room. A "move" consists of either admitting one person into the room or letting one out. Can you arrange a series of moves so that every possible combination of people is in the room exactly once?

- You've got an unlimited supply of bricks. You want to stack them, each brick overhanging the one beneath it. What is the maximum overhang you can create?

- You have to get from point A to point B. You don't know whether you can get there. What do you do?

- How do you find the closest pair of stars in the sky?

Ten

Weighing Your Head

What to Do When You Draw a Blank

In the 2006 Big Daddy Burger Eating Contest, held at Las Vegas's Plaza Hotel and Casino, contestants raced to devour a hamburger billed to be larger than David Hasselhoff's head. Some press agent invented that detail, banking on the theory that there's something intrinsically funny about David Hasselhoff. The problem was that no one knew how much Hasselhoff's head weighed. Probably Hasselhoff didn't know. "Unless we are certain of its weight, it is irresponsible to use Mr. Hasselhoff's head as a standard by which to measure this burger," said Richard Shea, president of the International Federation of Competitive Eating (IFCE) in an archly worded press release.

The IFCE invited Hasselhoff to participate in a nonsurgical weighing of his head. Hasselhoff's people took a pass, so the contest organizers arbitrarily set the burger's weight at nine pounds. Sonya "The Black Widow" Thomas won by wolfing one down in twenty-seven minutes.

This offbeat news story may be behind one of today's most diabolical interview questions: "How would you weigh your head?" The question has long been used in Oxford and Cambridge interviews. American technology companies seem to have begun asking it only in the past few years. Unlike the other riddles in this book, this one is incredibly hard and has no entirely satisfactory answer. It's mainly a way of putting the interviewee on the spot to see how he'll react to an impossible demand and almost certain failure.

As early as 1990, Joel Spolsky, a former Microsoft interviewer, tried asking, "How do they make M&M's?" Even he didn't know the answer. In a similar spirit is the Amazon tradition of a "bar raiser," a bad-cop interviewer who poses incredibly difficult questions outside the candidate's field of expertise. Amazon values generalists, but no one is a universal genius. Candidates are judged on how valiantly they struggle with the bar raiser's near-impossible questions.

That's becoming a common strategy in these desperate times for job seekers. Another notable example of the impossible question genre asks

? Can you swim faster through water or syrup?

It turns out that Sir Isaac Newton pondered this question more than three hundred years ago. Newton's answer was wrong. Fortunately, he never tried to get a job in Silicon Valley.

Salvaging a Doomed Interview

I have given many examples of how to answer tough questions in this book. Sooner or later, you'll meet up with a question you can't answer. It need not be as hard as the ones above. If you're stumped, you're stumped, and it's no consolation that some may find the question easy. There is, however, an art to salvaging an ill-fated response.

I'm not saying you can fake your way through these kinds of interview questions. I am saying it's better interview etiquette to keep trying to answer the question until the interviewer cuts you off. Interviewers ought to know that innovation takes persistence, intuition, and luck. You can at least show you've got the persistence part covered.

Your goal should be to avoid dead air. Silence makes both sides uncomfortable, and you'll be tempted to end the misery by announcing that you can't think of anything. In some interviewers' eyes, that makes you look like a quitter. Talking is also good because sometimes the brain follows the mouth. Rambling on about a problem may cause you to think of an approach you wouldn't have thought of otherwise.

The tough part is knowing what to say when you really don't see how to address the question. Here are some suggestions.

Restate the question. Put the question in your own words. This is a check on whether you've got the details right. It sometimes exposes gimmicky questions dependent on exact wording.

Disambiguate. No matter what the question is, you can ask for clarification or details. ("When you say I'm swimming in 'syrup,' do you mean a specific kind, like maple syrup, or any liquid that's thicker than water?") You'll get clues about what type of answer the interviewer expects. Asking for further information is vital with interactive questions in which it's meant for you to elicit the full details of the problem from the interviewer.

Describe why the obvious answer fails. These are difficult questions, and "difficult" generally means that the first answer that comes to mind won't be right. Adopting a skeptical tone, try to prove that the obvious answer or approach is wrong. ("I guess my first impulse is to say that I'd swim slower in syrup. But if it's thicker, there's more to push against....")

Analogize. Forming analogies—mental mappings—is key to all kinds of creative thinking. Describe ways in which the problem is comparable to something familiar—or something outlandish. ("The difference between swimming in syrup and swimming in water is something like the difference between flying a model airplane on the earth and flying one on the moon, where there's no air. What would happen to a model airplane on the moon?") No analogy is perfect, so you should also discuss how your analogies differ from the problem.

Brainstorm. With luck, your analogies will suggest other, not-so-obvious ways to tackle the problem. Try to list as many approaches as possible. Not all will be "good" answers, but that's okay. It's better to come up with lots of half-baked ideas than none at all. Furthermore, brainstorming is a way of trolling for hints. The most stony-faced interviewers may give some guidance, as long as they see you're trying.

Critique. You should analyze the problems with your brainstormed ideas. That usually gives you a lot to talk about, even when you're stumped, and may elicit more hints. Ideally, you would supply some closure by saying which approach is most promising.

Seesaws, Body Scans, Dead Ringers

I'll walk you through a possible answer to the "weighing your head" riddle. A restatement of the question might be, "You want me to weigh my head, using a scale of some kind, and without detaching it from my body?" This is correct.

There are several ambiguities to resolve by questioning the interviewer. One is, "What exactly do you mean by 'head'? Does the head include the neck, or only half the neck, or does it end immediately below the jaw?"

Another issue is whether you're allowed to use computations or estimates. "Does the scale actually have to register my head's correct weight, or can I make several measurements and compute the weight from them?"

Interviewers may vary on their answers to the first question. The point is, an ideal answer would let you draw the line between "head" and "body" anywhere you like. As to the second, this is not a Fermi question. You're not just estimating the weight of *a* head but determining the actual weight of *your* head. Computing is okay; guesstimating is not.

The obvious answer is (1) lie on the floor and put your head on a bathroom scale. Or recline on a chaise longue with your head extending beyond it, nestled in the tray of a butcher's scale set at a comfortable height. With the neck fully relaxed and the body otherwise supported, the scale would register the weight of your head and *only* your head.

You must have gathered that it's not that easy (though the IFCE suggested something like this to Hasselhoff). The next stage of your response might be to critique this answer. The head-on-scale approach *would* work, were the neck a limp string of negligible weight, connecting body and head. In reality, the neck has weight and cannot go completely limp (there's a spine inside). Posture and the tension of the neck could throw off the reading—or not. Who knows?

As an experiment, I tried this method. I lay on a carpeted floor, resting my head on a digital bathroom scale whose surface was about an inch above the carpet. The reading varied greatly with posture. Lying on my side produced readings averaging nine pounds. On my back (the most comfortable position), the average result was about eleven pounds. Facedown, with my head turned to the side, the reading was about seventeen pounds. In this position my face was pressed uncomfortably against the scale, and I could feel my shoulders transmitting weight.

I'm reasonably sure the eleven-pound reading is in the right area. I am also fairly confident that you will never encounter an interviewer who has tried this method, and for that reason you should steer clear of it. They won't believe it will work.

Moving on, the next step would normally be to brainstorm other possible approaches. Here are some common ones.

(2) The seesaw: Lie down on a seesaw, your neck just over the pivot and your body on the other side. The body side of the seesaw will be heavier, of course, and will touch the ground. Add weights to the head side until it just balances the body side. Figure it from that.

(3) The roundabout: A roundabout is the piece of playground equipment consisting of a big rotating disk, a kid-powered merry-go-round. Get on one, positioning your neck at the center of the disk. Have a friend set the roundabout spinning while making precise measurements of the force needed. Calculate your head's weight from the angular momentum.

(4) The body scan: Scan yourself with magnetic resonance imaging (MRI), which measures density of mass. Or use the new scanners at airports, which use low-level X-rays. Use the results to figure the percentage of weight in your head and find your answer from that.

(5) The corpse twin: Go to a medical school and find your body-double cadaver, as close as possible to your height, weight, and build. Persuade the management to let you weigh the cadaver, cut off its head, and weigh that, too. This will give the head's percentage of total body weight. Multiply that by your unclothed weight (that's assuming the corpse didn't have clothes on, either). If you find this too gruesome to contemplate, read further for a shortcut.

You can critique these answers, too, especially since none of them seems especially practical. With the seesaw and roundabout answers, the distance of weights from the pivot matters as much as the weights themselves. That's how a fat kid can balance a skinny kid on the seesaw—by sitting closer to the pivot. To calculate the weight of your head, you'd have to know the exact weight distribution of your body from head to foot. That's exactly what you don't know.

Both MRI and airport scanners provide X-ray vision (literally in the case of the airport machines), allowing the operators to see underneath clothes and spot anything denser than flesh. The MRI scan can measure the density of protons. That's hydrogen atoms in water, mainly, and the body is largely water. There are also hydrogen atoms in proteins and fats. Unfortunately, hydrogen atoms are not an exact proxy for body mass. Bone and teeth contain little hydrogen, and they're the densest materials in the body.

The body-scan approach probably sounds better than it is. Absent a NASA budget, would you really be able to look at the data and figure out the weight? How much work would that be, what would it cost, and when all is said and done, would it be any more accurate than faster, cheaper solutions?

Mulling over failed ideas can sometimes be informative. An issue that keeps coming up is that you don't know how heavy your head is, as a percentage of total weight. And that's about as far as you can go unless you have a *eureka!* moment.

Eureka! (or Not)

That's meant literally. The liberally educated will recall the story explaining the origins of the term *eureka.* King Hiero II of Syracuse commissioned a gold crown for a temple. He suspected that the goldsmith had cheated him by replacing some of the supplied gold with silver. Hiero asked Archimedes, his court technology adviser, to determine whether the cheaper metal had been substituted. This was way before chemistry existed. Everyone knew, however, that gold was denser than silver.

The crown had an irregular shape. It could not be placed in a balance with an equal volume of gold because no one could say what the equal volume was. Baffled, Archimedes went to the public baths. He noticed that the water level rose as he stepped in.

This gave him the solution. Archimedes leaped out of the bath and ran naked through the streets shouting, "I have found it!" ("Eureka!") Archimedes measured the volume of the crown by dunking it in water. That and the crown's weight allowed him to calculate its density. He demonstrated that the density was less than that of pure gold, proving the goldsmith was a crook.

Like many of the cool stories of antiquity, this may be fiction. It was related by Vitruvius, who lived a couple of centuries later. The incident is absent from Archimedes's own treatise *On Floating Bodies* and his other surviving works. True or not, Western culture has adopted this tale as a paradigm of how linear reasoning may thread the loopy complexities of the world.

You can draw an analogy between Archimedes's trick and the problem at hand. Both involve scales and measuring something of complex shape. Try this: (6) Set a bucket filled to the brim with water in a tray. Bend over and dunk your head in the bucket. Dunk far enough so that the waterline on your neck coincides with your imaginary dividing line between head and body. The dunking will spill water into the tray. Pull your head out and measure the water in the tray. It will equal the volume of your head.

The same trick, with a bigger container, allows you to measure the volume of your whole body. Then use a bathroom scale to measure your weight. Divide weight by volume to get the overall density of your body. Multiply by the head's volume to get its weight.

This calculation assumes that the head's density equals the body's density. Hmmm...you might ask the interviewer whether it's okay to assume that. But don't be surprised if he says no. The head's density isn't the same as the body's.

At this point, it's easy to get waylaid. The dunking measures your head's volume. There must be a way to get its density, you'd think. Hapless interviewees fill the whiteboard with equations and diagrams, devising tortuous, and torturous, experiments in buoyancy. These often involve upside-down weighings, with the body suspended by the feet, and the head underwater.

Save yourself the trouble. It can't be done. The buoyant force on a submerged object depends on the weight of the water displaced. (This is Archimedes's principle, mentioned in his treatise *On Floating Bodies.*) The weight of displaced water, in turn, equals the volume of whatever's doing the displacing (e.g., your head) times the density of water. It's the density of water, not of your head, that matters. It makes not the slightest difference whether your head is full of air — or something denser.

Unlike a typical brainteaser, this question doesn't have a magic answer where everything clicks into place. Answering it is more like solving a real-world problem. You have to make intelligent trade-offs and arrive at an outcome that's good enough for the purpose.

There are ways to refine your estimate of your head's density. The body is made of various substances with known densities. This is the basis of underwater weighing, used to measure the proportions of muscle and fat (and also ultimately based on Archimedes). The person sits in a seat attached to a spring scale. The seat is lowered into the water, and the scale measures how much lighter the person is when submerged. Standard formulas allow a clinician to determine the relative proportions of muscle and fat.

Air is the wild card of underwater weighing. There's a lot of air in the lungs and throughout the digestive tract. Almost all of this air is below the neck. Typically, the subject is instructed to exhale deeply before submersion. No one can exhale all the air in his lungs, and there's air in the stomach and intestines. This is estimated and subtracted using standard formulas.

By using these formulas, you can compute an air-free density of your body that will be greater — and more representative of your head — than a simple whole-body dunking would yield.

In fact, you might as well compute your muscle and fat

percentages, too, and use that to fine-tune the answer. Unusual physiques—those of bodybuilders, the anorexic, the obese—will be reflected in body density, but they shouldn't affect the head's density much. (A weight lifter doesn't add much muscle to his head.) In such cases you should adjust the density toward the average value and use that as the head's density.

You should always try to wrap up your answer. A good way to do that is to ask which of the methods you've described would be most accurate. If you thought of all the methods I've described, you might compare the merits of the two most promising approaches, Archimedes's and the corpse twin. In both cases, you start with your bathroom scale weight (easy and exact) and multiply by a ratio made from two rather exotic measurements. With the Archimedes answer, that's your head to body volume, derived from two dunkings, and with the corpse twin, it's head to body weight, from two weighings. Ideally, both of these ratios would be adjusted to allow for air and body build (Archimedes) or changes in weight distribution that occur after death and preservation (corpse twin). There's going to be guesswork in both adjustments. A reasonable hunch is that the corpse twin method is more accurate. Weighing a body that's not moving ought to be more accurate than a dunking experiment with a body that is moving and may have to come up for air before the water has settled.

Some will recall a line in the 1996 film *Jerry Maguire*, where Ray asks Jerry, "D'you know that the human head weighs 8 pounds?" This proves that you shouldn't get your facts from movies. Eight pounds would be light for an American adult of average build. Danny Yee, of the Department of Anatomy and Histology, University of Sydney, asked the dissection staff and got this answer: "An adult human cadaver head cut off around vertebra C3, with no hair, weighs somewhere between 4.5 and 5 kg, constituting around 8 percent of the whole body mass." Although you're not expected to know that percentage, you can tell the interviewer that it must be known to medical science and can be found by Googling—as indeed it can. If you weigh 150 pounds, your head weighs about 8 percent of that, or 12 pounds.

There Are No Rules

The "weighing your head" question verges on self-parody. Weighing the contents of your head is the purpose of today's immersive job interviews. It's a game where interviewer and interviewee both feel powerless, and the boundary between assessment and exploitation blurs.

Let me end with the most important single piece of advice in the book. You can improve your performance on tricky interview questions by embracing trial and error.

Most people don't. They imagine that the types of questions I've discussed in this book are all about logical deduction, or knowing arcane facts that happen to apply, or having some rare genius termed "creativity." They expect a more or less straight shot from question to answer. When instead they run into a dead end in their reasoning, they see that as a failure. Because of these false expectations, dead ends can have a disproportionate psychological effect, like awkwardly placed bunkers in a golf course. The interviewee freezes up and never recovers.

There's more evidence than ever that creativity entails trial and error. The new fashion in creativity research is today's ultimate form of head weighing, the MRI brain scan. Scanning appears to confirm the distinction between intelligence and creativity. Brains of people who score highly on psychological tests of creativity actually function more slowly in some situations than less creative people's brains. "The brain appears to be an efficient superhighway that gets you from Point A to Point B," explained Rex Jung of the Mind Research Network, Albuquerque. "But in the regions of the brain related to creativity, there appear to be lots of little side roads with interesting detours, and meandering little byways." Jung suspects that the slower nerve firing "might allow for the linkage of more disparate ideas, more novelty and more creativity."

Our minds are optimized to see obvious solutions to problems. Obvious solutions are generally right—except when they're not. Then the problem is considered "difficult." The questions in this book are all designed to be difficult by frustrating our reasonable hunches. You answer them by brainstorming not-so-obvious approaches and checking them to see which works best. Creativity can be a tough slog, and some consider it boring. So-called creative people are those who don't get bored or, at any rate, are motivated to keep plugging along. There is no magic algorithm for solving an unfamiliar problem. "Hell, there are no rules here," as Thomas Edison said. "We're trying to accomplish something."

Embedded in today's culture of interviewing is a similar tone of skepticism. It's a skepticism about established job descriptions, organization charts, industries, and even human relationships. Our ever-growing communications networks are engines of creative destruction, with the power to create and destroy business plans. It's an axiom that everything will be completely different in five years. There will be new rules, new ways of making money, new ways of living. This frantic ethos has employers deprecating fixed skills and fixed qualifications in favor of the seductive concept of mental flexibility. Sometimes you have to wonder, flexibility for *what?* How different can the world be in five years?

That's a puzzle we're all still working on. Meanwhile, those who succeed at today's interview mind games know how to take missteps in stride—to relax and find ways to enjoy the exploration of novel ideas. Perhaps success is a matter not so much of being smarter, but of being less entitled. And tenacity is a big part of creativity. That is the unstated thesis of today's interview by ordeal. As one former Google interviewer said, "The goal is to find out where the candidates run out of ideas."

Answers

With a standard logic puzzle, there's a right answer; not so much with many of today's most interesting interview questions. Where applicable, I point out both good and not-so-good answers and explain why there's a difference. There are often other good answers. If you've got a better one, go for it.

Chapter One

? You are shrunk to the height of a penny and thrown into a blender. Your mass is reduced so that your density is the same as usual. The blades start moving in sixty seconds. What do you do?

Those who were paying attention in rocket-science class will recall the formula for the energy of a projectile: $E = mgh$. E is energy (of a bottle rocket, let's say), m is its mass, g is the acceleration of gravity, and h is the height the bottle rocket attains. The height increases in direct proportion with energy (as long as mass stays the same).

Suppose you tape two bottle rockets together and light them

simultaneously. Will the double rocket go any higher? No; it's got twice the fuel energy but also twice the mass to lift against gravity. That leaves the height, h, unchanged. The same principle applies to shrunken humans jumping. As long as muscle energy and mass shrink in proportion, jump height should stay the same.

? When there's a wind blowing, does a round-trip by plane take more time, less time, or the same time?

The usual gut reaction is that the wind effects even out. A headwind will slow you one way; then it becomes a tailwind on the return and allows you to make up time. This is correct, as far as it goes. The question is, is the round-trip time *exactly* the same?

Say you've got a plane that travels from New York to London and back at 600 miles an hour. Coincidentally, a freak of global warming has generated a steady, 600 mph jet stream blowing from New York to London. This is great for the eastward trip. The hyperhurricane tailwind doubles the plane's ground speed and allows it to arrive in London in half the usual time.

It's the return trip that's a killer. The plane has a 600 mph headwind. No matter how much the pilot guns the engines, it can't overcome that. Even if the plane got in the air, its ground speed would be zero. ("Now, here, you see, it takes all the running you can do, to keep in the same place," the Red Queen told Alice in *Through the Looking-Glass.*) The plane will never make it back to New York. The "return trip" is of infinite duration, and so is the round-trip.

The predicament is easy to understand in this extreme case. On a 5-hour flight, a tailwind can save you (at the very most) 5 hours. A headwind can cost you an eternity. This basic principle is true no matter how fast or slow the wind is. A 300 mph wind would cut 1.67 hours off a 5-hour flight one way, but add 5 hours the other way. A constant wind always increases the round-trip time.

To make your answer complete, you might also talk about a

crosswind. Suppose the wind is blowing from the north, at a ninety-degree angle to the route from New York to London. Should you ignore the wind and steer the usual course, the wind would push the plane south throughout the trip. It would end up somewhere south of London. To correct for the sideways wind, you need to chart a course slightly north of London, *into* the wind. This means that some of the plane's velocity fights the wind, leaving a reduced component for eastward travel. The trip takes a little longer. On the return, you've got the same crosswind and have to correct the same way. *Both* trips take longer.

In general, you wouldn't expect the wind to be blowing exactly in the direction of travel, nor exactly at ninety degrees to it. The direction would be somewhere in between. You can break the wind's velocity down into headwind-tailwind and crosswind components. The point is that both components increase the round-trip time. The best wind for round-trip air travelers is no wind at all.

? What comes next in the following series?

SSS, SCC, C, SC

The series is the letters of the alphabet in a silly code. *A*, as a capital letter, is made of three straight lines. Encode that as SSS. Capital *B* is one straight line and two curved ones, or SCC. *C* is one curved line and remains, coincidentally, just C.

D is one straight and one curved line. That brings us up to the next term, which must represent a capital *E*. That's four straight lines, or SSSS.

This question has become popular at Amazon, but it's probably too insight dependent to be a good test. Many smart people think of binary numbers, roman numerals, and other concepts that will get them absolutely nowhere.

? You and your neighbor are holding garage sales on the same day. Both of you plan to sell the exact same item. You plan to put your item on sale for £100. The neighbor has informed you that he's going to put his on sale for £40. The items are in identical condition. What do you do, assuming you're not on especially friendly terms with this neighbor?

The bit about not being on especially friendly terms should tell you that a strategic response is expected. So should the fact that this question is frequently asked at the more aggressive Wall Street houses.

Figure that your time is worth something. You've got better things to do with your weekends than to run a perpetual series of garage sales. Therefore you try to price items so that the probability of their selling is high. You want to get rid of nearly everything by day's end.

Assuming that the same applies to the neighbor, there must be an honest difference of opinion on how much the item in question is worth. Provided that at least one person is willing to pay £40, the neighbor can count on his item selling, decreasing the chance that your item will. Your worry is that there will be only one "big spender" willing to pay £100, and he'll buy from the neighbor, not from you.

The friendly solution is to pull the neighbor aside and say, "Now look, a mint-condition Wookiee in original packaging is worth a hundred pounds. You can check it out on eBay. You're throwing money away by offering it for forty pounds." This may persuade the neighbor to raise his price, maybe matching your £100. But this plan is not considered an especially good answer. Suppose the big spender finds two identical items on sale for £100.

He's equally likely to choose either one, and the other may go unsold.

Ironically, you might be better off with the neighbor *lowering* his price. Were the neighbor to give the item away to the first person who showed up, then you wouldn't have to worry about it anymore. You simply want the neighbor's item off the market, one way or another.

You might offer to pay the neighbor *not* to market his item. It's hard to say whether he'd accept. He might be offended, and he might try to extort an unreasonable price. A better answer is simpler: buy the neighbor's item.

Why? First of all, he'll be pleased to sell his item immediately. He's not likely to be offended or to raise the price. You can haggle, like any other buyer, and may get it for less than £40.

Why should you want his item? When you put something on sale for £100, you hope to make a decent profit, compensating for the time you've invested in selling it and factoring in the chance that it won't sell. Anything that diminishes the chance of your item's selling in effect costs you a significant fraction of that £100. The numbers in this puzzle were chosen so that the neighbor's price is comparable to the economic damage he's doing to you. By buying the item, you get the right to keep it off the market, when that suits your purposes, *plus* the right to sell it at any price the market will bear. Anything you get from selling the second item is pure gravy.

The best plan is to hide one item until the first one sells. Then put the second item on sale at a reduced price, according to how late in the day it is.

? You put a glass of water on a record turntable and begin increasing the speed slowly. What will happen first: will the glass slide off, will it tip over, or will the water splash out?

This question is asked at Apple. Most people understand

that the question involves centrifugal force. Equally important is the force of friction. It's friction between the bottom of the glass and the turntable that sets the glass in motion.

To make that clear, imagine a world without friction. Everything is slicker than Teflon, infinitely slick. Then the question's experiment would have no effect on the glass. The turntable would slide effortlessly underneath the glass, and the glass wouldn't move. This is in keeping with Newton's first law: stationary objects stay put unless there's some force acting on them. Without the force of friction, the glass won't spin.

Now imagine super-gluing the glass to the turntable, effectively creating infinite friction between the two surfaces. Glass and turntable must rotate as a unit. Crank up the speed, and the glass moves faster. This creates centrifugal force. The only thing free to react to the force is the water. It's not glued down. Once the glass is spinning fast enough, the water will spill out of the outward side, away from the center of rotation.

The question asks you to consider a case between these extremes. At first, friction will be adequate to hold the glass in place. It will rotate with the turntable, creating a slight centrifugal force. As the turntable spins faster, the centrifugal force will increase. The friction holding the glass in place will remain approximately the same. Therefore, there must come a point when the centrifugal force overcomes the frictional force.

Those who have studied physics or spent a lot of time on playground slides will recall that an object that begins to slide experiences less friction than it did while standing still. You "stick" a little at the top of the slide, then suddenly slip freely. The same applies to the turntable. Rather than accelerating gradually, the glass will stick, then slide.

What happens then? The answer is, *it depends on the shape of the glass and how full it is.* This isn't a weaselly attempt to evade the question. All the following outcomes are realistic:

1. Fill a glass to the brim. The slightest centrifugal force will raise the water level on the cup's outward side, spilling some water. This will occur while the glass is "sticking"—before it slides.
2. Use a very short "glass," a petri dish with a single drop of water in it. No way is the glass going to tip over, nor will it be moving so fast that the single drop of water climbs the surface and spills out. Instead, when it's spinning fast enough, the petri dish–glass will simply slide off the turntable.
3. Use a very tall glass, like a test tube with a flat bottom. The centrifugal force effectively acts on the center of gravity. Because the center of gravity is so high, and all the frictional force is on the very bottom, the test tube–glass will flip over rather than slide.

The turntable surface makes a difference, too. A rubber turntable increases friction and favors spilling and tipping scenarios, all else being equal. A slick, hard plastic turntable favors sliding.

Chapter Two

? It is difficult to remember what you read, especially after many years. How would you address this?

In a few years, practically all reading will be on digital screens of some kind. It's possible to envision a personal reading agent keeping track of everything you've read, on any device—from e-mails and tweets to electronic books and magazines. This agent or its data would migrate from hardware device to device and follow you throughout life.

• The agent would index all that text so that you'd be able to do a keyword search.

• The agent would let you annotate things you read anywhere (as you can with e-book readers). The simplest annotation would be a highlight, essentially meaning "I may want to remember this later!" The highlighted passages should get extra weight in searches. You should be able to search the text of annotations, too.

• This agent could become a built-in function of Google search or its future counterparts. While Google "remembers" that newspaper article you read last October, and (via Google Books) everything in many books, the profusion of hits in a given search can be overwhelming. If Google (via this agent) kept track of what you read, on any and all devices, you could filter searches by "stuff I've read."

• You can search for something you *remember* you've forgotten, but not for something you've *completely* forgotten or aren't thinking about. Maybe you read the complete short stories of Nikolay Gogol at school and loved them but hardly remember a character, plot, or phrase now. The agent might address that by periodically reminding you of highlighted passages, those you've identified as worth the attention of your future self. Perhaps the agent would have a Twitter account and tweet random snippets of material you want to remember but don't otherwise have time to go back to.

• The agent would also "remember" podcasts, movies, and TV shows—provided transcripts or subtitles exist or could be generated.

? There are three men and three lions on one side of a river. You need to carry them all to the other side, using a single boat that can carry only two entities (human or lion) at a time. You can't let the lions outnumber the men on either bank of the river because then they'd eat them. How would you get them across?

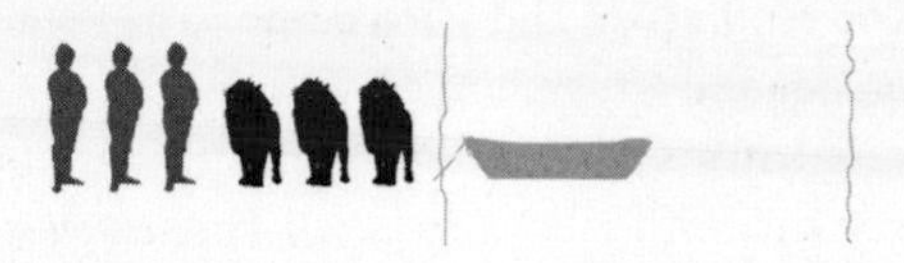

There are five conceivable passenger lists for the first trip, namely, one man; one lion; a man and a lion; two men; or two lions.

Lions can't row or trim sails. (You'd be surprised how many people miss this point.) A boat without a man onboard is a non-starter. That rules out *one lion* and *two lions.*

One man and *two men* are out, too: they would leave the man or men outnumbered on the near bank.

That leaves only one workable case for the first trip, a man and lion crew. They travel to the far shore.

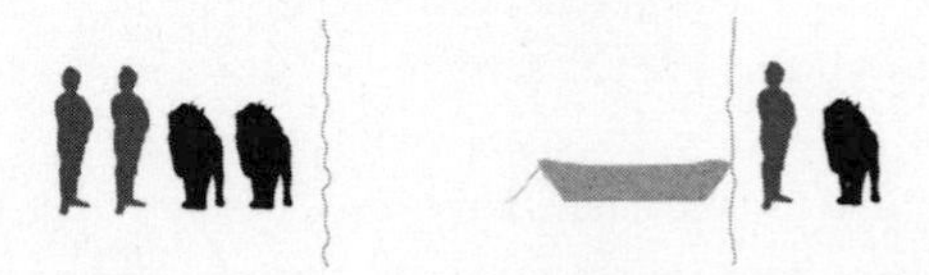

Now what? Nothing's going to happen until we get the boat back to the near bank. It's not going to sail itself. The lion's not going to sail it. That means the man must return alone. This gives us:

Now look at the options for the next trip. You can't send two men across because that would leave one outnumbered. The only safe possibilities are *man and lion* or a lone *man.* But sending a lone man would be pointless. He'd just have to turn around and come back. So send a man and a lion across. The man will have to push the lion out and return immediately, for otherwise he would be on the same bank with two lions.

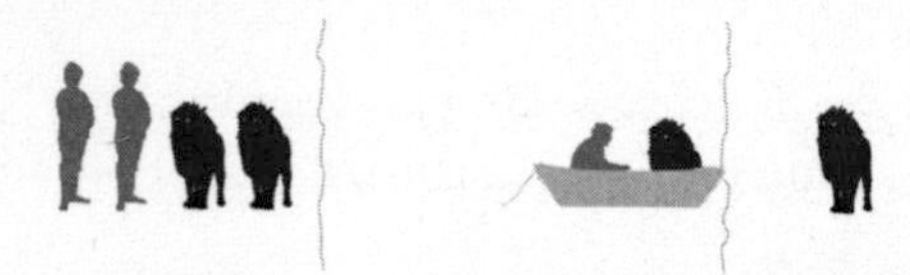

That leaves two lions on the far bank. Once the man returns, all three men are on the near bank, along with a lion.

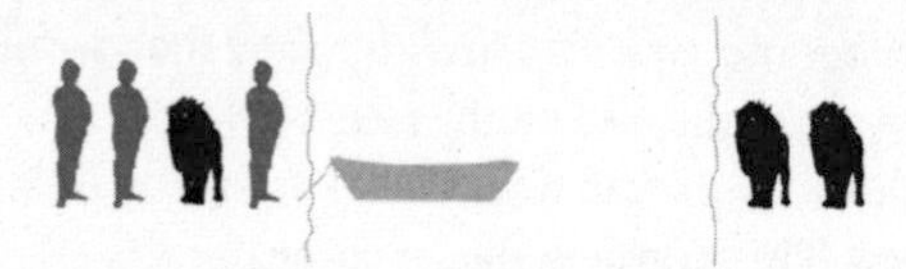

So far every action has been forced. The next trip presents an authentic choice. We can send *two men* or *a man and a lion.*

In the latter case, the man would again have to drop off the lion and return immediately. This would create a toxic lion majority. With three lions on the far shore, there would be no safe way of *ever* landing a man or men there again. Scratch that.

Instead, send two men to the far shore. Since they're not outnumbered by the two lions there, they can get out of the boat and stretch their legs.

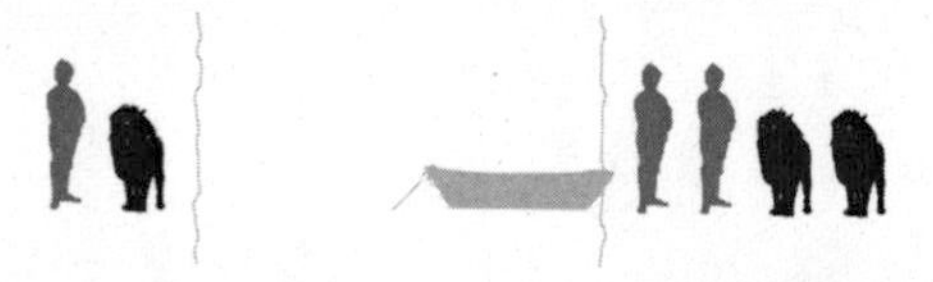

The return has to be either *two men* (a pointless undoing of the previous trip) or *a man and a lion.* It can't be just one man because that would leave the other outnumbered. So a man and a lion return to the near shore.

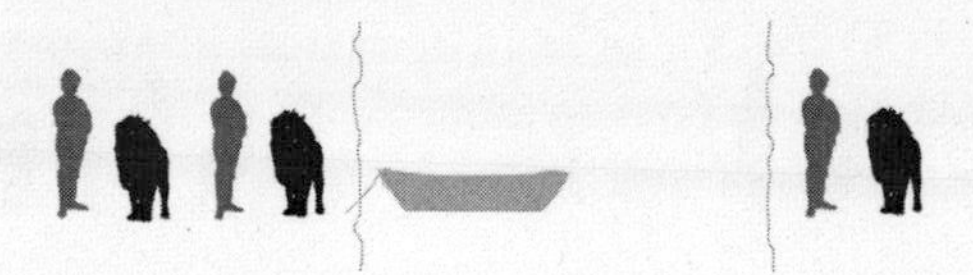

For the trip after that, we don't want to waste our time by sending the man and the lion right back. The only safe alternative is to send two men:

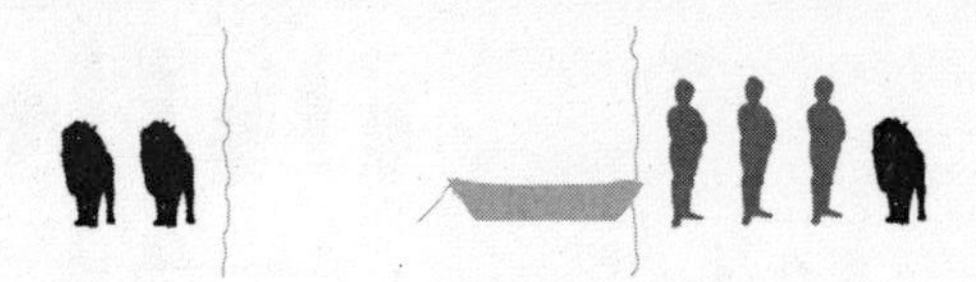

Send a single man back to pick up a lion. He can't get out, as he'd be outnumbered two to one. He's got to grab a lion or coax it to jump into the boat. A slab of meat would be helpful.

Once he's got the lion in the boat, the man returns with it to the far shore.

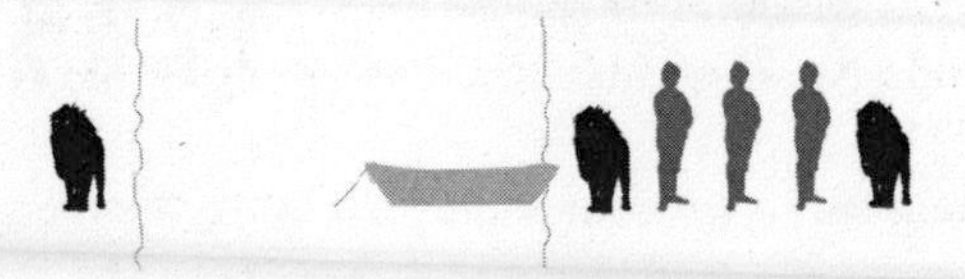

Then the man goes back for the remaining lion. This time he can get out on land if desired.

Finally, man and lion return to the far shore. That's every mammal across in five and a half round-trips.

This puzzle is a politically correct update of one that played a role in early artificial intelligence research. In 1957, Allen Newell and J. Clifford Shaw of the Rand Corporation and Herbert Simon of the Carnegie Institute of Technology unveiled the General Problem Solver, one of the first AI problems. Its creators had asked people to solve logic puzzles while narrating their reasoning. The computer scientists then boiled down the techniques and coded them into the General Problem Solver. One of the test problems involved three cannibals and three missionaries crossing a river. Aside from the cast, it's identical to this interview puzzle.

In the broadest sense, these puzzles are much older than that. The eighth-century Alcuin of York wrote a famous puzzle collection, *Propositiones ad acuendos juvenes (Problems to Sharpen the Young),* including one about a man crossing a river with a wolf, a goat, and a basket of cabbages. It was probably old even then.

? Using only a 4-minute hourglass and a 7-minute hourglass, measure exactly 9 minutes.

With a 4-minute hourglass, it's a cinch to measure 4 minutes, 8 minutes, 12 minutes, and so forth. The 7-minute hourglass readily gives multiples of 7. You can measure still other times by "adding" the two hourglasses—starting one the instant the other finishes. Let the 4-minute glass run out, then start the 7-minute glass. This gives 11 minutes total. Similar strategies measure 15 minutes (4 + 4 + 7), 18 minutes (4 + 7 + 7), and so on.

This method *won't* measure 9 minutes. But there's another trick, "subtraction." Start both hourglasses simultaneously. The instant the 4-minute glass runs out, turn the 7-minute glass on its side to stop the sand. There is then 3 minutes' worth of sand in one bulb. That sounds promising. Nine is 3 × 3. But notice that once you "use" the 3 minutes of sand, it's gone. You end up with all 7 minutes of sand back in one bulb. You could repeat the whole process twice, but that doesn't permit measuring a continuous 9 minutes.

The way to remedy this is a third trick that might be called "cloning." Start both glasses at 0 minutes. When the 4-minute glass runs out, flip it over. When the 7-minute glass runs out, flip both glasses over. The 4-minute glass, which had 1 minute left to run, now has 3 minutes after the flip. When those 3 minutes run out, flip over the 7-minute glass again. It will then have 3 minutes of sand. (You've "cloned" the 3 minutes that were in the small glass.) This gives a continuous 9 minutes.

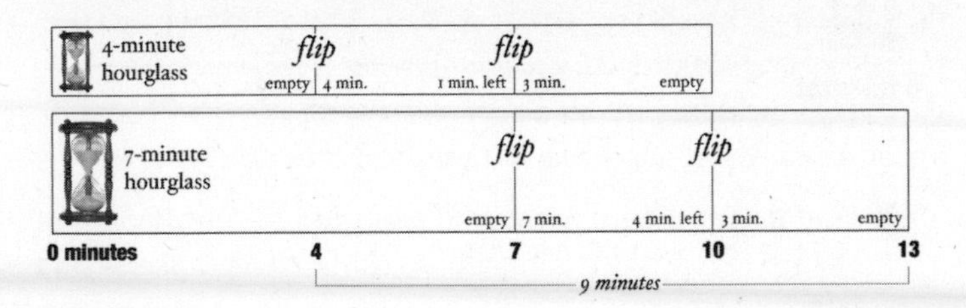

The above is a good answer, just not the best one. Its defect is that it requires 4 minutes of preparation time (to get 3 minutes of sand in one bulb of the 7-minute glass). The scheme therefore takes 13 minutes total, in order to measure 9 minutes. Would you buy an egg timer that took 4 minutes to warm up?

There is a solution that allows the timing to begin immediately. Begin with the safe assumption that we're going to start both hourglasses at 0 minutes. Then fast-forward to 7 minutes later. The 7-minute hourglass has just run out. The 4-minute glass has already run out once and (presumably) has been flipped over. It should have just 1 minute of sand remaining in its upper bulb.

All that's needed is to clone that 1 minute. At 7 minutes, flip over the 7-minute glass. Let it run 1 minute, as measured by the remaining sand in the 4-minute glass. That brings us up to the time 8 minutes. The 7-minute glass will have 1 minute of sand in its lower bulb. Flip the 7-minute glass over again and let the minute of sand run back. When the last grain falls, that will be 9 minutes.

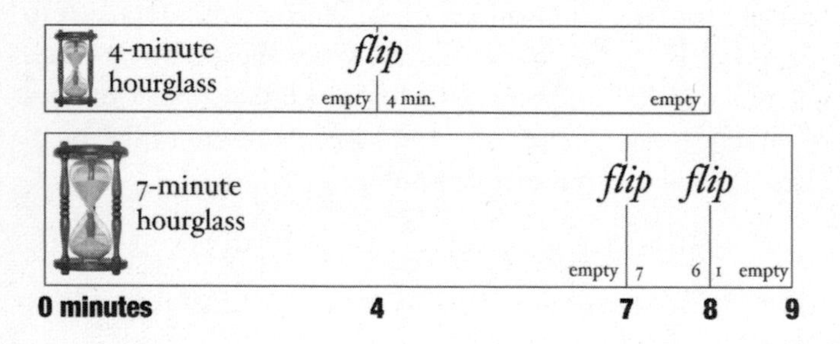

? Find the minimum number of coins to give any amount of change. (From a U.S. company, so think 1¢, 5¢, 10¢, 25¢ and 50¢.)

There are two ways of interpreting the question. They have

different answers, so you should ask the interviewer which is meant. One interpretation is, find the smallest assortment of coins that will give exact change in any needed amount, from 1 cent to 99 cents. Call this a universal change-making set. How many coins are in that set?

Say you're a finicky shopkeeper who likes to start the day with just enough coins in the till so that you will be able to make change on the first day's transaction, no matter what. What is the smallest set of coins that will do that?

The answer is easy because the denominations of American coins were chosen to facilitate making change. Every denomination is at least twice the value of the next-smaller denomination. That means you can use this algorithm for making X cents' change:

> If the needed amount X is 50 cents or more, put down a 50-cent piece and subtract that from X;
>
> If X is now 25 cents or more, put down a quarter and subtract that from X;
>
> Divide X by ten and take the whole part. Put down that many dimes and subtract;
>
> If the remainder is 5 cents or more, put down a nickel and subtract;
>
> Divide the remainder by 1 cent and put down that many pennies.

Not only does this rule work, but it gives change with the fewest coins. You could, for instance, skip the first line and use two quarters rather than one 50-cent piece, but that would mean using an extra coin.

Someone wanting to make any amount of change in the

fewest coins will need one 50-cent piece, one quarter, and one nickel and will never need more than one of each. Two dimes might be required (to make 20 cents, say), and up to four pennies (to make 4 cents). That means there are nine coins with a combined value of $1.04 in a universal change-making set. Obviously, you never use all nine to make change from a dollar.

An alternate interpretation of the question is, *what's the smallest number,* X, *such that you will never need more than* X *coins to make change?* This essentially asks you to consider what change amount requires the most coins. You might guess that 99 cents requires the most coins, and you'd be right. It takes eight coins, namely a 50-cent piece, a quarter, two dimes, and 4 cents. Making 94 cents' change also requires eight coins (replace one of the dimes with a nickel).

This question is considered tricky enough to appear on psychological tests of creativity.

? In a dark room, you're handed a deck of cards with *N* of the cards faceup and the rest facedown. You can't see the cards. How would you split the cards into two piles, with the same number of faceup cards in each pile?

This brainteaser is popular at JP Morgan Chase. Nowadays, it's reasonable to say you would pull out your mobile phone and use the screen as a flashlight. Actually, the puzzle predates mobile phones and is solvable without seeing the cards. You'd probably begin by making these observations:

- Arbitrarily dividing the deck into two equal piles won't work (unless you're very lucky). All the faceup cards could be in one pile.
- The question doesn't say the two piles have to be equal—only that they contain the same number of faceup cards.

- You can flip cards over. Of course, you have no way of telling whether the cards you're flipping are faceup or facedown.

The intended answer is that you count N cards off the top of the deck and flip them over. That's one pile. The remainder of the deck is your second pile.

Here's why it works. The N cards you counted off could have any number of faceup cards, from zero to all N of them. Let's say there were f faceup cards (before the flip). Flipping converts every faceup card to a facedown card and vice versa. So instead of f faceup cards, you end up with $N - f$ faceup cards in this pile.

The other pile, the remainder of the deck, has N faceup cards minus the f you counted off. That's the same amount as in the flipped pile.

? You're given a cube of cheese and a knife. How many straight cuts of the knife do you need to divide the cheese into twenty-seven little cubes?

To create twenty-seven little cubes, you have to cut the original cube into three slices in each of three directions. It takes two cuts to produce three slices. The obvious answer is to make two slices parallel to each of the three axes, or six slices total.

With this kind of question, the first answer that pops into your head usually isn't the best one. Can you do better? You're allowed to reposition the pieces after each slice (as cooks often do when dicing onions). This greatly increases the number of possibilities, and you may find your spatial intuition lagging.

But there is actually no way to do this in fewer than six slices. Ideally, you should prove that to the interviewer. Here's how. Picture the innermost cubelet after you've sliced the original cube into $3 \times 3 \times 3 = 27$ pieces. This cubelet has no outside surface.

Therefore you've got to create each of its six faces with a knife slice. Six straight slices is the bare minimum for doing that. This is a reverse trick question. The obvious answer is right, and many founder trying to devise a nonobvious one.

Martin Gardner identified this puzzle's author as Frank Hawthorne, a supervisor with New York's Department of Education, who published it in 1950. The notion of rearranging the cut pieces to save slices is not crazy. You can cut a cube into $4 \times 4 \times 4$ cubelets in just six slices (versus the nine needed with a simple slice and dice).

In 1958, Eugene Putzer and R. W. Lowen published a general solution for optimal slicing of a cube into $N \times N \times N$ cubelets. They assured any practical-minded readers that their method might "have important applications in the cheese and sugarloaf industries."

This question loosely recalls another posed in interviews at some financial firms: How many cubes are in the center of a Rubik's Cube? Since the standard cube is $3 \times 3 \times 3$, the fake-out answer is "one." Anyone who's ever disassembled a Rubik's Cube knows the real answer is "zero." There's a spherical joint in the middle, no cubelet.

? There are three boxes, and one contains a valuable prize; the other two are empty. You're given your choice of a box, but you aren't told whether it contains the prize. Instead, one of the boxes you didn't pick is opened and is shown to be empty. You're allowed to keep the box you originally picked ("stay") or swap it for the other unopened box ("switch"). Which would you rather do, stay or switch?

This interview question is a version of the Monty Hall dilemma, posed by biostatistician Steve Selvin in 1975. Monty Hall was the original host of the TV game show *Let's Make a Deal.* Selvin's puzzle concerns a situation loosely based on the TV show's final round, in which contestants chose prizes behind doors. In a letter to the *American Statistician,* Selvin argued that you should switch boxes, an answer so controversial that Selvin had to defend it in a follow-up letter. Monty Hall himself wrote Selvin, agreeing with his analysis.

Since then the dilemma has been the subject of interminable debate. It became popular with the general public after appearing in a 1990 letter to the *Parade* magazine columnist Marilyn vos Savant. The following year, the *New York Times'* John Tierney reported that the puzzle "has been debated in the halls of the Central Intelligence Agency and the barracks of fighter pilots in the Persian Gulf. It has been analyzed by mathematicians at the Massachusetts Institute of Technology and computer programmers at Los Alamos National Laboratory...." The puzzle has turned up on the radio show *Car Talk* and the television show *NUMB3RS.* It's used in interviews at Bank of America and other financial firms. Cynics might detect a parallel to the financial industry's risk management, in which probabilities are invisibly shifted and someone else is left holding the empty box.

The most interesting thing about Selvin's puzzle is how hard it is. One study found that only 12 percent of those questioned gave the right answer. That's amazing when you consider that someone who's completely clueless can guess and stand a fifty-fifty chance of being right! This is a case where intuition gives a bum steer.

The majority opinion is that it makes no difference whether you keep your original box or switch. The more sophisticated may add that any who think they can better their expectations by switching are as misguided as the losing slot-machine player who insists the machine is "overdue" to hit the jackpot.

With any question in probability, it is important to know which parts of the story happen by chance and which by design. Say a friend tosses a coin ten times, and it comes up heads each time. What is the chance that the next toss will come up heads, too? You can't say until you find out whether the run of heads is a freak of luck or the result of a trick coin.

When Selvin posed the puzzle, the original *Let's Make a Deal* was still on the air and an ingrained part of pop culture. My grandmother, who watched the show, regarded Monty as a glorified con man. Her reasoning—recited loudly to the TV set—was that "if he's willing to offer you that door, he must know it's less valuable than what you've already got."

She wasn't far off the mark. In interviews, Hall has said that when he knew a contestant had chosen the biggest prize, he would offer cash incentives to tempt the contestant to switch. It made better TV to see a sucker trade away a big prize for junk.

Call the three boxes the chosen box, the revealed box, and the temptation box. Initially, the chance that your chosen box contains the prize is one in three.

Now one of the two remaining boxes is opened and revealed to be empty. In order to determine how this affects the odds, you need to know who chooses this second box and with what agenda. There are two likely cases:

1. The revealed box is chosen randomly—say, by a coin toss—from the two boxes you didn't choose. This means that the revealed box might have contained the prize, though as it happened, it didn't.
2. The box is chosen by someone who knows what's in the boxes and who planned all along to reveal an empty box—*which he can do no matter what.*

Selvin's original puzzle makes it perfectly clear that the second case is the intended interpretation. ("Certainly Monty Hall knows which box is the winner and therefore would not open the box containing the keys to the car.")

This all-important clarification is often omitted in the retelling. As stated, the interview puzzle is ambiguous. There's no mention of a trickster-emcee, no indication of how the revealed box is chosen. You should ask the interviewer for details and point out that the question has different answers, depending on how the second box is chosen.

In the first case, opening the revealed box tells you something. It reveals that the prize isn't in that box, even though it could have been. That boosts the chance of the chosen box's containing the prize from 1/3 to 1/2. It has the same effect on the odds of the temptation box. Since both now have a fifty-fifty chance of containing the prize, there's no point in switching.

In the second case, revealing one box doesn't tell you anything meaningful. Monty (or whoever) knows what's in the boxes and can always find an empty box to show you. His scripted revelation does not budge the chance that your original box is the winner. It was 1/3 initially, and it is still 1/3.

Opening the revealed box has not changed the 2/3 probability that *one of the other two boxes* contains the prize, either. But since one of these boxes has already been shown to be empty, the 2/3 probability must now reside entirely in the temptation box. By accepting the offer to switch to it, you *double* your chances of getting the prize.

If you still have trouble seeing why Selvin's answer is right, imagine that there are a hundred boxes. You choose box #79. Monty then opens 98 of the 99 other boxes. All are empty. That leaves, besides your box, box #18, say. Monty asks if you want to swap box #79 for #18.

You started with 99-to-1 odds against there being the car keys in your box. Monty's actions are just stage business. He has no intention of showing you anything except an empty box and is fully able to do so. The chance of the prize's being in your box remains fixed at 1/100 while the chance of its being in box #18 increases to 99/100. With 100 boxes, you would increase your chances 99-fold by switching.

When the psychologists Donald Granberg and Thad A. Brown interviewed people presented with this dilemma, they kept hearing explanations like these:

> "I wouldn't want to pick the other door because if I was wrong I would be more pissed off than if I stayed with the 2nd door and lost."
>
> "It was my first instinctive choice and if I was wrong, oh well. But if I switched and was wrong it would be that much worse."
>
> "I would really regret it if I switched and lost. It's best to stay with your first choice."

These are expressions of loss aversion. It's universal human nature to recoil from a decision that might leave one worse off, even when the odds are favorable. "Better safe than sorry." Anyone who invents new products would do well to keep this in mind. The consumer thinking of switching boxes or brands may be motivated by reasons that have nothing to do with logic.

Maths geniuses are as loss averse as everyone else. It's said that the famed mathematician Paul Erdös got this puzzle wrong the first time he heard it. "Even Nobel physicists systematically give the wrong answer," said the psychologist Massimo Piattelli-Palmarini, "and... they insist on it, and they are ready to berate in print those who propose the right answer."

? You're in a car with a helium balloon tied to the floor. The windows are closed. When you step on the accelerator, what happens to the balloon—does it move forward, move backward, or stay put?

The near-universal intuition is that the balloon leans backward as you accelerate. Well, the intuition is wrong. Your job is to deduce how the balloon does move and to explain it to the interviewer.

One good response is to draw an analogy to a spirit level. For the not so handy, a spirit level is the little gizmo carpenters use to make sure a surface is horizontal. It contains a narrow glass tube of colored liquid with a bubble in it. Whenever the spirit level rests on a perfectly horizontal surface, the bubble hovers in the middle of the tube. When the surface isn't so level, the bubble migrates to the higher end of the tube. The takeaway here is that the bubble is simply a "hole" in the liquid. When the surface isn't level, gravity pulls the liquid toward the lower end. This pushes the bubble wherever the liquid *isn't*—toward the opposite end.

Untie the helium balloon and let it hit the sunroof. It becomes a spirit level. The balloon is a "bubble" of lower-density helium in higher-density air, all sealed in a container (the car). Gravity pulls the heavy air downward, forcing the light balloon against the sunroof.

When the car accelerates, the air is pushed backward, just as your body is. This sends a lighter-than-air balloon forward. When the car brakes suddenly, the air piles up in front of the windshield. This sends the balloon backward. Centrifugal force pushes the air away from the turn and sends the balloon toward the center of the turn. Of course, the same applies when the balloon is tied to something; it's just less free to move. The short answer to this question is that the balloon nods in the direction of any acceleration.

Don't believe it? Put the book down right now. Go to the supermarket, buy a helium balloon, and tie the string to the gear stick or hand brake. Drive back home (no lead-footing necessary). You'll be astonished. The balloon does exactly the opposite of what you'd expect. When you step on the accelerator, it bobs forward, like it's trying to race the car to the next light. Brake hard enough to throw the kids' toys out of the backseat, and the balloon pulls backward. In a high-speed turn, as your body leans outward, the crazy balloon veers inward. It's so freaky that there are videos of this on YouTube.

Why are our intuitions right about spirit levels and wrong about helium balloons? In a spirit level, the heavy liquid is dyed a fluorescent sports-drink hue, while the bubble is a ghostly void. We associate color with density, and transparency with nothingness. That instinct is completely wrong with balloons. Air is invisible, and we ignore it 99+ percent of the time. The balloon, on the other hand, is dressed up in pretty colors or Mylar and screams, "Look at me!" We forget that, masswise, it's a partial vacuum within the surrounding air. A helium balloon does the opposite of what a mass does because it's a deficit of mass. The real mass—the air—is invisible.

Interviewers who ask this question don't expect you to know much physics. But there is an alternate response that makes use of the theory of relativity. Seriously.

It relates to Albert Einstein's famous thought experiment about the lift. Imagine you're in a lift going to your tax accountant's office, and a mischievous extraterrestrial decides it would be fun to teleport you and the lift into intergalactic space. The lift is sealed, so there's enough air inside to keep you alive long enough to amuse the alien for a few minutes. There are no windows, so you can't look out and see where you are. The alien puts the lift in a tractor beam and tows it at a constant acceleration exactly matching that of Earth's gravity. Is there anything you could do in the sealed lift to determine whether you're experiencing real Earth gravity or "fake" gravity, mimicked by acceleration?

Einstein said no. Should you take your keys out of your

pocket and drop them, they would accelerate toward the lift floor exactly as they would on Earth. Let go of a helium balloon's string, and it would float upward, just as on Earth. Things would appear perfectly normal.

The Einstein equivalence principle says that there is no (simple) physics experiment that can distinguish between gravity and acceleration. This assumption is the foundation of Einstein's theory of gravity, known as general relativity. Physicists have been trying to punch holes in the equivalence principle for nearly a century now. They haven't been able to. It's safe to assume that Einstein's premise is right, at least for any experiments you can do in a car with a fifty-pee balloon.

All right, here's a physics experiment. Tie the string of a plumb bob (carpenter's weight) to your right index finger. Tie a helium balloon to the same finger. Note the angle between the two strings.

In a lift, a parked car, or a cruising jetliner, the outcome will be the same. The plumb bob points straight down. The balloon points straight up. The two strings, joined at your finger, form a straight line. This is the outcome whenever you are subjected to gravity.

Now picture what happens when you start driving. As you accelerate, your body sinks back in the seat. Fallible intuition may tell you that the plumb bob and the balloon will each lean backward a bit from your finger. During acceleration, there will be an angle between the two strings (*if* this intuition is right). This would provide a way of distinguishing between gravity and acceleration. When the car is subject to gravity alone, the two strings form a straight line. But when it's subject to centrifugal force or other forms of acceleration, the strings form an angle with your finger as the vertex. That's all you need to prove that general relativity is wrong. Forget about getting a job at Google—this would be worthy of a Nobel Prize.

But since the equivalence principle has been rigorously tested and shown to be true, this doesn't happen, and you can use it to answer this question. Physics must be the same in an accelerating car as in a car subject to gravity alone. In both cases, the balloon, your finger, and the plumb bob form a straight line. In answer to the question, then, the helium balloon does the exact opposite of what you'd expect of an object with mass. It goes forward rather than backward...left rather than right...and, of course, up rather than down.

Chapter Three

> **?** According to a survey, 70 percent of the public likes coffee, and 80 percent likes tea. What are the upper and lower bounds of people who like both coffee and tea?

Not all tea drinkers like coffee; not all cat lovers like dogs; not all QPR fans are Chelsea fans. Draw a Venn diagram on the whiteboard or in your mind. It will have a rectangle representing the totality of survey participants. Let most of the rectangle represent

the 70 percent of the public that likes coffee, and draw a small circle to indicate the 30 percent of the public that evidently dislikes coffee. (The regions must add up to 100 percent, though the areas aren't necessarily in proportion.)

Eighty percent of the public likes tea. If this percentage were drawn as a circle, it would have to overlap both the coffee-drinker and coffee-hater regions. (There just aren't enough coffee drinkers to account for all the tea drinkers.) To set an upper bound on the people who like both beverages, assume that every coffee drinker likes tea. The circle representing the 80 percent of tea drinkers would therefore be split among those who like both tea and coffee (70 percent) and those who like just tea (10 percent). The 70 percent is the upper bound.

To get the lower bound, slide the tea-drinker circle over so that it engulfs the coffee-hater circle. Now everyone who dislikes coffee (30 percent) likes tea. That leaves $80 - 30 = 50$ percent who like tea and also like coffee. That's the lower bound.

? At 3:15, what is the angle between the minute and hour hands on an analog clock?

It's not zero. At 3:15 the minute hand will be pointing due east, at the 3. The hour hand will already have moved one-quarter of the way from 3 to 4. The span between 3 and 4 is one-twelfth of a full 360-degree turn, or 30 degrees. Divide that by four and you've got the answer, 7.5 degrees.

? How many integers between 1 and 1,000 contain a 3?

Some numbers (like 333) contain more than one 3. You don't want to count them twice (or three times). The question is asking how many different numbers contain *at least one* 3.

Every number from 300 through 399 contains at least one 3. That's a hundred numbers right there.

There are also a hundred numbers with 3 in the tens place—30 through 39; 130 through 139; up to 930 through 939. We've already counted a tenth of them, namely the numbers 330 through 339. That's ten numbers to scratch from the tally. This leaves 100 + 90 = 190 numbers so far.

Finally, there are a hundred numbers ending in 3, from 3 to 993. Scratch the ten that begin with 3 (303, 313, 323,...,393). That leaves ninety. One-tenth of that ninety have a 3 in the tens place (33, 133, 233,...,933). Scratch those nine, leaving eighty-one.

The grand total is $100 + 90 + 81 = 271$.

? A book has N pages, numbered the usual way, from 1 to N. The total number of digits in the page numbers is 1,095. How many pages does the book have?

Every page number has a digit in the units column. With N pages, that's N digits right there.

All but the first 9 pages have a digit in the tens column. That's $N - 9$ more digits.

All but the first 99 pages have a digit in the hundreds column (accounting for $N - 99$ more digits).

I could go on, but not many books have more than 999 pages. A book with 1,095 digits in its page numbers won't, anyway.

This means that 1,095 must equal

$$N + (N - 9) + (N - 99)$$

This can be simplified to

$$1{,}095 = 3N - 108$$

That means that $3N = 1{,}203$, or $N = 401$. That's the answer, 401 pages.

? How many 0s are there at the end of 100 factorial?

One hundred factorial—written "100!"—is 100 multiplied by every natural number smaller than itself. It looks like this:

$$100 \times 99 \times 98 \times 97 \times \cdots \times 4 \times 3 \times 2 \times 1$$

For this question, you're not supposed to multiply 100! out. You're expected to deduce how many 0s are at the end of the product without knowing that product.

To accomplish that, you'll have to formulate some rules. You already know one of them. Look at this equation:

$$387{,}000 \times 12{,}900 = 5{,}027{,}131{,}727$$

See something funny? When you multiply two round numbers with 0s on the end, you can't possibly get an unround number, with no 0s on the end. That violates the law of conservation of ending 0s (a law that I just made up, but which is true enough). A product always inherits the ending 0s of its factors. Some correct examples of this:

$10 \times 10 = 100$

$7 \times 20 = 140$

$30 \times 400 = 12{,}000$

Of the factors of 100!, ten end in a 0. They are 10, 20, 30, 40, 50, 60, 70, 80, 90, and 100 (which ends in two 0s). That makes eleven ending 0s in the factors, which 100! necessarily inherits.

Warning: this leads some unfortunate interviewees to give eleven as their answer. Wrong. Sometimes you can multiply two 0-free numbers and get a product that has 0s. Examples are

$2 \times 5 = 10$

$5 \times 8 = 40$

$6 \times 15 = 90$

$8 \times 125 = 1{,}000$

All but the last pair occur in the hundred factors of 100! There's still some work to do. This brings us to the law of hot dogs and buns. At a barbecue, some people bring hot dogs (in packs of ten), some bring buns (in packs of eight), and some bring both hot dogs and buns. There's only one way to figure out how many complete sandwiches you can serve. Count the hot dogs, count the buns, and take the smaller number of the two.

The same law answers this interview question, once you replace "hot dogs" and "buns" with "factors of 2" and "factors of 5."

In each of the equations I just gave, a number that divides by 2 is being multiplied by a number that divides by 5. The 2 and 5 factors "join" to produce a perfect 10, which adds a 0 to the product. Look at the last example, where three 0s appear out of thin air:

$$8 \times 125 = (2 \times 2 \times 2) \times (5 \times 5 \times 5)$$

$$= (2 \times 5) \times (2 \times 5) \times (2 \times 5)$$

$$= 10 \times 10 \times 10$$

$$= 1{,}000$$

It's all about pairing up 2s and 5s. Take a number like 692,978,456,718,000,000. It's got six 0s on the end. That means it can be written this way:

$$692{,}978{,}456{,}718 \times 10 \times 10 \times 10 \times 10 \times 10 \times 10$$

or this way:

$$692{,}978{,}456{,}718 \times (2 \times 5) \times (2 \times 5) \times (2 \times 5) \times (2 \times 5) \times (2 \times 5) \times (2 \times 5)$$

The first term, 692,978,456,718, does *not* divide by 10. If it did, it would end in 0, and we would already have factored out another 10. As it is, there are six factors of 10 (or 2×5), corresponding to six 0s at the end of 692,978,456,718,000,000. Reasonable enough?

That yields a foolproof system for determining how many 0s are at the end of any large number. Factor the number by 2s and by 5s. Pair the factors off, $(2\times5) \times (2\times5) \times (2\times5) \times \cdots$. The number of matching 2 and 5 *pairs* equals the number of 0s on the end. Ignore any leftovers.

In general you are going to have some 2s or 5s left over. It's usually the 2s. In fact, it's *always* the 2s when you're dealing with a factorial. (The factorial will contain more even factors than factors divisible by 5.) It's therefore the number of 5s that is the bottleneck. The question becomes, how many times can 100! be divided evenly by 5?

That's easy to tally in your head. From 1 through 100, there are twenty numbers that divide by 5: 5, 10, 15,. . . , 95, 100. Notice that 25 adds two factors of 5 ($25 = 5 \times 5$) to the product, and so do the three multiples of 25, namely 50, 75, and 100. That adds four more 5s, for a total of 24. The 24 factors of 5 pair up with an equal number of 2s, producing 24 factors of 10 (and leaving a lot of leftover 2s). There are therefore twenty-four 0s on the end of 100!

In case you're curious, the exact value of 100! is

93,326,215,443,944,152,681,699,238,856,266,700,490,715,
968,264,381,621,468,592,963,895,217,599,993,229,915,608,941,
463,976,156,518,286,253,697,920,827,223,758,251,185,210,916,
864,000,000,000,000,000,000,000,000

Chapter Four

? Explain the significance of "dead beef."

"Dead beef" is hexspeak. During debugging, the contents of computer memory must be displayed on a screen (or on paper in the old days). That could be rendered as a completely unreadable sea of 0s and 1s. A postage-stamp-size sample of it might look like this:

00000001

10000101

10010101

00100010

Instead, it's conventional to display computer memory in hexadecimal notation. That's base 16, using the standard digits 0 through 9 plus the letters A, B, C, D, E, and F (standing for the numbers ordinary mortals call 10, 11, 12, 13, 14, and 15). The result is more concise, though still hard to make sense of.

B290023F

72C70014

993DE110

8A01D329

Coders have always been desperate to create recognizable landmarks within that alphanumeric morass. They realized that certain hexadecimal numbers look like English words in strident uppercase. It's possible to "spell" any word or phrase using the first six letters of the alphabet (and sometimes using 0 for the letter *O* and 1 for *I* or lowercase *l*). Examples: FEEDFACE, ABADBABE, DEADBABE, and, you guessed it, DEADBEEF.

0993FF10

7229B236

22C74290

DEADBEEF

Some IBM and Apple Mac systems periodically wrote DEADBEEF to memory. This made it easy to tell whether memory had been corrupted by buggy code. If you didn't see the expected DEADBEEF values, you knew something was seriously wrong.

DEADBEEF is not a universal code, and many other hexspeak words are used for that and other purposes. The question basically tests whether the candidate is close enough to the culture of computing to have heard of it.

? There's a latency problem in South Africa. Diagnose it.

"Latency problem in South Africa" is an inside joke at Google. The phrase is intentionally equivocal techspeak, like a line in science fiction. ("We're losing potency in our antimatter pods!") The candidate should be able to figure out what it *might* mean, though, and say something sensible.

"Latency" means a delay. That could apply to almost anything, from getting a marriage license to using public transit. It's a reasonable guess that a Google interviewer is thinking about the Internet. The interviewer could mean either of the following:

- The Internet is running slowly in South Africa.
- Google searches (only) are running slowly.

The ping operation measures latency on the Internet. A ping is a dummy message sent from point A to point B and back. The time interval is a measure of how fast information is flowing. By pinging from many computers and stations in South Africa, you can tell whether the Internet infrastructure is slow there. If not, the problem may be with Google. Are there enough servers for the South African traffic? Try a set of search terms from many points in South Africa, to see whether all are slow or just some. This would allow you to map the (imaginary) problem, and that generally satisfies the interviewer.

? Design an evacuation plan for San Francisco.

The 2006 *Emergency Evacuation Report Card* of the American Highway Users Alliance gave Kansas City an A grade. New Orleans, reeling from Katrina, got a D. San Francisco's grade? F. New York, Chicago, and Los Angeles failed, too.

The failing grades are due to these cities' size, hemmed-in geography, and dependence on public transportation. At such an environmentally conscious place as Google, some interviewees instinctively talk up San Francisco's public transit network. But most public transit runs within the urban area. (BART, the San Francisco underground, can get people to Oakland. Is that good enough? Or are we evacuating Oakland, too?) Amtrak doesn't even stop in San Francisco proper. For the foreseeable future, there's no such thing as a green evacuation. Emptying a city on short notice means internal combustion engines on public roads.

Here are some bullet points to incorporate into your plan:

- Make use of the fact that everyone wants to get out of the city as fast as possible. Allow a marketplace of transportation options. The biggest hitch in the Katrina evacuation was that New Orleans authorities couldn't issue timely traffic advisories: they simply didn't know which roads were jammed. Katrina hit the year before Twitter and a couple of years before ubiquitous smartphones. Your plan should encourage people to tweet or text about traffic conditions (but not while driving!) and should devise a way to swiftly incorporate this information into social networks, mapping applications, broadcast media, and so forth.

• Use school buses. The USA's school buses have greater capacity than all the modes of adult "mass transit" combined. Organize free school bus shuttle service for those who don't have cars.

• Divert petrol to the region's petrol stations. There were fuel shortages in the Katrina evacuation.

• In an authentic emergency, most people can't leave fast enough, but you have to worry about three classes of stragglers: those who refuse to go; those who can't evacuate without help (they're disabled or in hospitals); and those so off the grid that they won't hear about the evacuation (probably, many of them homeless or elderly). As a legal and practical matter, there's not much that can be done when a resident chooses to stay behind. Efforts are better spent canvassing neighborhoods for people who want to evacuate but need help. Put into service all the existing dial-a-ride vans and ambulances, as they have special facilities for the frail and disabled.

• Make sure some of the buses and trains allow pets and suitcases. One reason people insist on staying behind is concern for their pets and valuables.

• Designate all lanes of traffic arteries outbound. This allows twice as much traffic and prevents the clueless from entering the city. Known as contraflow, this idea will be familiar to Bay Area commuters. Since 1963, the Golden Gate Bridge has had reversible lanes. In the mornings, four of the six lanes are inbound to San Francisco. The rest of the time, there are three lanes each way, to the city and the Marin County suburbs.

? Imagine a country where all the parents want to have a boy. Every family keeps having children until they have a boy; then they stop. What is the proportion of boys to girls in this country?

Ignore multiple births, infertile couples, and couples who die before having a boy. The first thing to realize is that every family in the country has, or will have, exactly one boy when they're done procreating. Why? Because every couple has children until they have a boy, and then they stop. Barring multiple births, "a boy" means one boy exactly. There are as many boys as completed families.

A family can have any number of girls, though. A good way to proceed is to take an imaginary census of girl children. Invite every mother in the country to one big room and ask on the public-address system: "Will everyone whose first child was a girl please raise her hand?"

Naturally, one-half of the women will raise a hand. With *N* mothers, *N*/2 would raise their hands, representing that many firstborn girls. Mark that on the imaginary tote board: *N*/2.

Then ask: "Will everyone whose second child was a girl please raise—or keep raised—her hand?"

Half the hands will go down, and no new hands will go up. (The mothers whose hands were down for the first question, because their first child was a boy, would not have had a second child.) This leaves *N*/4 hands in the air, meaning there are *N*/4 second-born girls. Put that on the tote board.

"Will everyone whose third child was a girl raise or keep raised her hand?" You get the idea. Keep this up until finally there are no hands still up. The number of hands will halve with each question. This produces the familiar series

$$(1/2 + 1/4 + 1/8 + 1/16 + 1/32 + \cdots)\times N$$

The infinite series sums to 1 ($\times N$). The number of girls equals the number of families (N) equals the number of boys (or very close to it). The requested proportion of boys to girls is therefore 1 to 1. It's an even split after all.

? On a deserted road, the probability of observing a car during a thirty-minute period is 95 percent. What is the chance of observing a car in a ten-minute period?

This question is challenging only because the information it supplies isn't the information you want. That's the way real life works.

You want to derive a ten-minute probability from a thirty-minute one. You can't just divide 95 percent by three (not that some haven't tried). It isn't particularly enlightening to know the chance of a car's passing in thirty minutes because that could happen in many ways. A car could pass in the first ten-minute segment, or the second, or the third. Two cars could pass, or five, or a thousand, and that would still count as "a car" passing.

What you really want to know is the probability that *no car will pass* in a thirty-minute period. That's simple. Since there's a 95 percent chance of at least one car passing in thirty minutes, there must be a 5 percent chance of no car in that time frame.

In order to have a thirty-minute stretch with no cars whatsoever, three things must happen (or rather, not happen). First, ten minutes must pass with no cars. Then, another ten minutes must pass, still without a car. Finally, a third ten minutes must be car-free. The question asks for the chance of a car's passing during a ten-minute period. Call that chance *X*. The chance of no car in ten minutes is 1 − *X*. Multiply that by itself three times, and it should come to 5 percent:

$$(1 - X)^3 = 0.05$$

Take the cube root of both sides:

$$1 - X = \sqrt[3]{0.005}$$

Solve for *X*:

$$X = 1 - \sqrt[3]{0.005}$$

Nobody expects you to do cube roots in your head. A laptop will tell you the answer comes to about 63 percent. That makes sense. The chance of a car in a ten-minute period should be less than the 95 percent chance of one in a thirty-minute period.

? You have a choice of two wagers: One, you're given a basketball and have one chance to sink it for £1,000. Two, you have to make two out of three shots, for the same £1,000. Which do you prefer?

Call the probability of making a basket *p*. With the first bet, you have a *p* chance of winning £1,000. Otherwise, you get nothing. On the average, you can expect to win £1,000 × *p*.

With the second wager, you shoot three times and have to make the basket twice to be in the money. The chance of making the basket on any given attempt is still *p*. Your chance of missing on any attempt is 1 − *p*.

There are 2^3, or 8, scenarios for the second wager. Let's list them (as you may do on a whiteboard during the interview). A check mark means you make the shot; a blank means you miss.

1st shot	*2nd shot*	*3rd shot*	*Probability*	*Win the £1,000?*
			$(1-p)^3$	No
		√	$p(1-p)^2$	No
	√		$p(1-p)^2$	No
	√	√	$p^2(1-p)$	Yes
√			$p(1-p)^2$	No
√		√	$p^2(1-p)$	Yes
√	√		$p^2(1-p)$	Yes
√	√	√	p^3	Yes

The first scenario is the one where you're really off your game. You miss all three shots. The chance of that is $1 - p$, multiplied by itself three times. You don't get the money.

In four of the eight scenarios, you win the money. In three of them you miss once. These scenarios have the probability of $p^2(1 - p)$. In the case where you make all three shots, the probability is p^3. Add all of them up. Three times $p^2(1 - p)$ comes to $3p^2 - 3p^3$. Add p^3 to that, to get $3p^2 - 2p^3$. The expectation is £1,000 × $(3p^2 - 2p^3)$.

So which wager is best?

First wager's expectation: £1,000 × p
Second wager's expectation: £1,000 × $(3p^2 - 2p^3)$

You may be a complete klutz (p is roughly 0) or an NBA baller (p approaches 1). For reference, I've done what you can't do in the interview: plugged the formulas into a worksheet and made a chart. The chart shows how expected winnings vary with p.

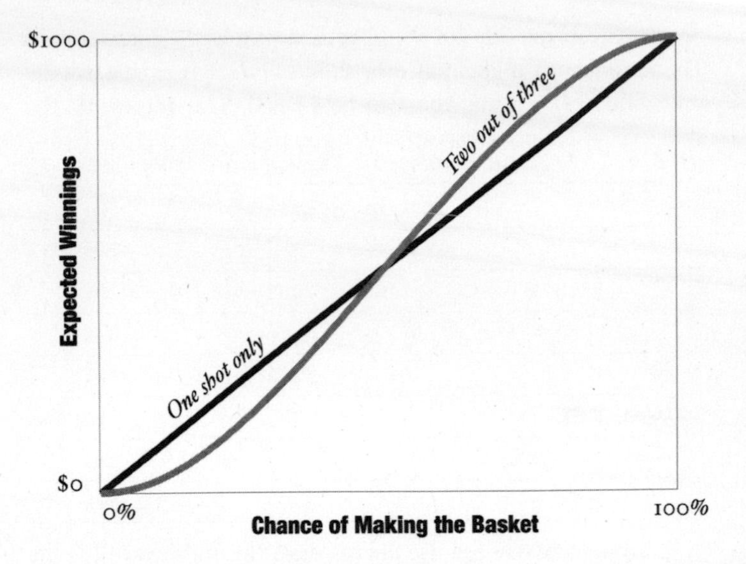

The straight diagonal line represents the first bet, and the more S-shaped curve is the second. The first bet is better if your chance of making the shot is less than 50 percent. Otherwise, you're better off picking the second wager.

This makes sense. A poor player cannot expect to win either wager. He must pin his hopes on a freak lucky shot, which is obviously more likely to happen once than twice ("lightning never strikes twice"). The bad player is better off with wager 1.

The very good player ought to win either wager, though there is a small chance he'll miss the money shot. Two out of three is a better gauge of his talents, and that's what he wants. It's like the legal maxim: if you're guilty, you want a jury trial (because anything can happen); if you're innocent, you want a judge.

* * *

Assuming you get this far, the interviewer's follow-up question is, "What value of p makes you switch bets?" To answer that, set the probability of winning the two bets equal. This represents the skill level where it's a toss-up which you pick.

$$p = 3p^2 - 2p^3$$

Divide by p:

$$1 = 3p - 2p^2$$

and then get

$$2p^2 - 3p + 1 = 0$$

From there you can use the quadratic formula, warming the heart of your old algebra teacher. The interviewer will be looking for brio as much as book learning. You know p, a probability, has to be between 0 and 1. It's better style to experimentally try a reasonable value. "Okay, I need a number between 0 and 1. Let's try 0.5." Works like a charm.

? Use a programming language to describe a chicken.

In 1968, the French writer and prankster Noël Arnaud published a slim volume of poems in the computer language ALGOL (now obsolete, it was a precursor of C). Arnaud restricted himself to ALGOL's short dictionary of twenty-four predefined words. The poems were not valid code. Describing a chicken in ALGOL, or C++, could be an exercise in the same quixotic spirit.

Interviewers usually intend for you to describe an *individual* chicken so that it might be distinguished from other members of its species. Pretend you're starting a social network site for poultry. "The chicken named Blinky is female, friendly, and dead." They want something like that, in legal code or pseudocode.

One example that would satisfy most interviewers:

```
class Chicken
{
public:
bool isfemale, isfriendly, isfryer, isconceptualart,
isdead;
};
int main()
{
Chicken Blinky;
Blinky.isfemale = true;
Blinky.isfriendly = true;
Blinky.isfryer = true;
Blinky.isconceptualart = true;
Blinky.isdead = true;
}
```

? There's a staircase and you're allowed to ascend one or two steps at a time. How many ways are there to reach the *N*th step?

Start simple. You're standing on the landing and want to reach the first step, #1. There's just one way to do it—take one step up.

Now let $N = 2$. There are two ways to get to the second step. Either you take two single steps in succession or you take one double step.

This is practically all you need to solve the problem. To see why, imagine your goal is step #3. For the first time, you can't get there in a single bound. It's got to be a combination of steps. But there are only two ways of arriving at step #3—either by taking a single step (from step #2) or by taking a double step (from step #1). We already know that there is only one way to get from the landing to step #1. We also know that there are only two ways to get from the landing to step #2. Add these ($1 + 2 = 3$) to get the number of ways to arrive at step #3.

The same logic applies to every higher step. There are two ways to get to step #4, from step #2 or from step #3. Add the number of ways of getting to step #2 (2) to the number of ways of getting to step #3 (3). That gives 5, the number of ways of getting to step #4.

It's easy to continue the series. The number of ways to climb quickly snowballs, looking like this:

Step:	1	2	3	4	5	6	7
Ways:	1	2	3	5	8	13	21

For anyone with a maths background, the lower series will look familiar. It's the Fibonacci sequence. (More on that in a moment.) The interviewer wants an answer for the general case of N stairs. It's simply the Nth Fibonacci number.

Leonardo Fibonacci, also known as Leonardo Pisano, was the most influential Italian mathematician of the late Middle Ages. It was Fibonacci who realized the incredible superiority of the Arab-Hindu system of numeration, with its place notation, over the roman numerals still used in medieval Europe. With the Arab-Hindu system, multiplication and division could be reduced to an algorithm (another Arab word). With roman numerals, these operations were impractically hard. Merchants had to seek out expensive experts to perform calculations on an abacus. In 1202, Fibonacci wrote a guide to the abacus, *Liber abaci*, in which he pitches "Arabic" numerals to what must have been a skeptical audience. The book also describes what we now call the Fibonacci sequence. Fibonacci didn't invent it; the sequence was known to sixth-century Indian scholars.

Start by writing 1, then another 1 after it. Add them to get the sum (2) and append that sum to the series.

1 1 2

To generate each new number, just add the last two numbers in the series. The series becomes

1 1 2 3 5 8 13 21 34 55 89 144...

Conspiracy theorists will find that the Fibonacci series turns up in all sorts of unexpected places. Want to convert between miles and kilometers? Use adjacent Fibonacci numbers (55 miles per hour = 89 kilometers per hour). The next time you've got too much time on your hands, count the little fruitlets making up a pineapple. You'll find that they form two intersecting sets of helices running in opposite directions. One set has eight helices; the other, thirteen. Both are Fibonacci numbers. Similar patterns are seen with pinecones, sunflowers, and artichokes. Coincidence? Not likely, nor is the fact that the Fibonacci sequence turns up in *The Da Vinci Code* (it's a combination to a safe)—and in this interview question, used at an information company bent on world domination.

? You have N companies and want to merge them into one big company. How many different ways are there to do it?

In the proper sense of "merger," two companies surrender their identities and fuse into a brand-new entity. The pharmaceutical giants Glaxo Wellcome and SmithKline Beecham merged in 2000 to form the pharma colossus GlaxoSmithKline. (You guessed it—both parent companies were themselves merger spawn.)

CEO egos being what they are, true mergers are fairly uncommon. Mergers require a near-exact match of bargaining power. More typically, one company's management has the stronger hand, and it's not about to let the weaker company's leaders forget it. The deal is likely to be an acquisition, in which company A gulps up company B, and B ceases to exist as a separate entity (though it often survives as a brand). An example is Google's 2006 acquisition of YouTube.

Mergers are symmetrical. There is only one way for two companies to merge as equals. Acquisitions are asymmetrical. There are two ways for two companies to acquire or be acquired—Google buying YouTube is *not* the same as YouTube buying Google.

Most people outside investment banking gloss over the distinction between mergers and acquisitions. Any melding of corporations is loosely called a "merger." The point is, you need to ask the interviewer what is meant by "merge." Fortunately, most of the reasoning applies whatever the interviewer's answer.

Start with acquisitions because they're more common (and a little easier to work with). You can visualize the companies as draughts, and the acquisitions as the moves in the game. Start with N pieces. A move consists of putting one piece on top of another to signify that the top piece is "acquiring" the bottom piece. After an acquisition, you manipulate the pieces involved as if they were glued together (like a "kinged" piece in the regular game).

Every move diminishes the number of pieces (or stacks) by one. Eventually, you will be placing stacks of pieces on top of other stacks to create yet-taller stacks. It will take exactly $N - 1$ moves to achieve the game's goal, a single tall stack consisting of all N pieces combined into one. How many different scenarios can lead to that outcome?

The simplest case involves two companies. Company A can gulp up B, or B can gulp up A. That's two possible scenarios.

With three companies, you have to decide which company first acquires what other company. There are six possibilities for that first acquisition, corresponding to the six possible ordered pairs of three items (AB, AC, BA, BC, CA, and CB). After the initial acquisition, you're left with two companies. The situation is then exactly as in the paragraph above. The number of possible acquisition histories for three companies is therefore $6 \times 2 = 12$.

With four companies, there are twelve possibilities for the first acquisition: AB, AC, AD, BA, BC, BD, CA, CB, CD, DA, DB, and DC. That decided, you have three companies and, as we already know, twelve histories. There must be $12 \times 6 \times 2$, or 144, acquisition histories for four companies.

Let's generalize. With N companies, the number of possible initial acquisitions is

$$N(N - 1)$$

This just means that any of the N companies can be the first acquirer, and any of the remaining $N - 1$ companies can be the first acquiree. After the first acquisition, there will remain $N - 1$ distinct companies, and there will be $(N - 1)(N - 2)$ possibilities for the second acquisition. Then there will be $(N - 2)$ companies and $(N - 2)(N - 3)$ possible acquisitions. We're going to be multiplying the ever-decreasing numbers of possible acquisitions until we're left with 2×1 possibilities for the final acquisition.

It's easy to see that, using factorial notation, the product will come to $N! \times (N - 1)!$ acquisition histories.

What if you want true mergers instead of acquisitions? The above analysis overcounts the possibilities by a factor of two—for each of the $N - 1$ mergers. That means the number of proper merger histories is $N! \times (N - 1)!$ divided by 2^{N-1}.

Finally, if "merger" can mean merger *or* acquisition, you simply add the two answers.

? What is the most beautiful equation you have ever seen? Explain.

Asked of Google engineers, this question prompts you to analyze how an equation can be "beautiful" and then to give a suitable example. The one certain thing about beauty is that it's subjective. Even so, most conclude that a beautiful equation is concise and of universal significance. Note, however, that you're not just trying to think of a beautiful equation; you're trying to impress the interviewer with your originality. It helps to give an equation that the interviewer doesn't hear every day.

Most would agree this is a lame answer:

$$E = mc^2$$

It's like a politician saying his favorite movie is *Titanic*.

You want Einstein? A better reply is

$$\boldsymbol{G} = 8\pi \boldsymbol{T}$$

This packs the general theory of relativity into five characters. **G** is the Einstein tensor, representing the curvature of space-time. **T** is the stress-energy tensor, measuring the density of mass and energy. The equation says that mass-energy curves space and time (which curvature we experience as gravity).

Another five characters express much of quantum physics.

$$\hat{H}\Psi = E\Psi$$

This is Schrödinger's equation, read as "the Hamiltonian of the wave function equals its energy."

The canonical Google answer is Euler's equation. It connects five numbers central to mathematics: *e*, pi, the imaginary number *i*—and of course 1 and 0, which are pretty important in the information business.

$$e^{\pi i} + 1 = 0$$

Euler's equation is regularly voted the "most beautiful equation" or something of the kind. It was tied for first (with all four Maxwell's equations!) in a 2004 *Physics World* poll for the "greatest equations ever." As one reader put it, "What could be more mystical than an imaginary number interacting with real numbers to produce nothing?"

"Like a Shakespearean sonnet that captures the very essence of love, or a painting that brings out the beauty of the human form that is far more than just skin deep, Euler's equation reaches down into the very depths of existence," the Stanford mathematician Keith Devlin wrote. The best-known commentary of all is probably that of Carl Friedrich Gauss, who said that unless this formula was immediately obvious to a student, that student would never be a first-rate mathematician.

You won't gain any points for originality by answering with Euler's equation, though. It's like saying your favorite film is *Citizen Kane.*

The Gaussian integral has some of the same mystic appeal, connecting *e*, pi, and infinity. One point in its favor: Gauss didn't find it completely obvious.

$$\int_{-\infty}^{\infty} e^{-x^2}\,dx = \sqrt{\pi}$$

The Gaussian integral also has something that Euler's equation lacks—relevance to life as we live it. The e^{-x^2} is the Gaussian function. A chart of it is the familiar bell-shaped curve of a normal probability distribution. This is the "curve" that teachers grade on—the one that supposedly governs heights, IQ scores, and the random walk of stock prices (but doesn't quite). The Gaussian blur filter in Photoshop uses the same function to blur your ex out of the picture.

In the equation, the integral computes the area under the bell-shaped curve and finds it equal to the square root of pi, or about 1.77. The equation can be viewed as a symbol of the role of chance in the world. Many of the things we value most—beauty, talent, money—are the result of scores of random factors, ranging from genes to simple luck. When the factors determining a quantity are truly random and additive, the quantity will follow a normal distribution. Most people will be in the middle of the curve. A few outliers will have a lot more or a lot less than the mean. In 1886, Francis Galton said of this distribution,

> I know of scarcely anything so apt to impress the imagination as the wonderful form of cosmic order expressed by the "law of error." A savage, if he could understand it, would worship it as a god.... Let a large sample of chaotic elements be taken and marshalled in order of their magnitudes, and then, however wildly irregular they appeared, an unexpected and most beautiful form of regularity proves to have been present all along.

For the cult of the beautiful equation in all its delirious glory, look no further than the British physicist Paul A. M. Dirac. "It is more important to have beauty in one's equations than to have them fit experiment," he once wrote. Dirac was notoriously eccentric and socially awkward, partly the result of autism. As a theoretical physicist, he saw the world as a puzzle to which beautiful equations were the key. To a remarkable degree, contemporary science (and many job interviewers) accept Dirac's view of the world.

For an amusing rebuttal, see Richard Feynman in the second volume of *The Feynman Lectures on Physics,* in which he makes the amazing claim that all of physics can be reduced to a single equation. The equation is

$$U = 0$$

That's it! That's everything about the universe!

Feynman was half-serious. Take an equation like $E = mc^2$. It is said to be *deep.* Its so-called beauty rests with the fact that it explains so much with just a few marks on paper, a few black pixels on white. This perception of simplicity rests on hard-won and messy concepts, Feynman argued. What is energy? What is mass? What is the speed of light? None of these concepts existed for al-Khwarizmi or Leonardo da Vinci. Energy and mass had only started to gel by Newton's time. "The speed of light" was hardly a scientific matter until the nineteenth century. Feynman's point is that $E = mc^2$ is an *abbreviation.* Admire it, just don't get wrapped up in how "simple" it is. It's not all that simple.

Notice that you can transform Einstein's equation into

$$E - mc^2 = 0$$

All I've done is to subtract mc^2 from both sides. They were equal before, and they must be equal now. Now square both sides of the equation. This gives

$$(E - mc^2)^2 = 0$$

The point of this will become clear in a moment. It's part of Feynman's recipe for the ultimate beautiful equation. Blend in a few more equations. For the heck of it, we'll use Schrödinger's equation and Euler's equation. Leave Euler's as is and tweak Schrödinger's

$$e^{\pi i} + 1 = 0$$

$$\hat{H}\Psi - E\Psi = 0$$

Then square both sides of each and add to Einstein's rejiggered equation

$$(E - mc^2)^2 + (\hat{H}\Psi - E\Psi)^2 + (e^{\pi i} + 1)^2 = 0$$

All three terms on the left have to be 0 (say Einstein, Schrödinger, and Euler). The equation must be correct, assuming its three components are. Furthermore, the only way the equation can be correct is if all the terms are 0. (That's the point of the squaring. It guarantees that no term could be negative. The only way for three nonnegative terms to add up to 0 is for all to be 0.)

Don't stop there. Throw in the kitchen sink, said Feynman. All equations, great and trivial, can be put in this form and appended to the left side of this master equation. Feynman called the quantities on the left $U_{N,}$ where N ranges from 1 to as far as you care to go. Sum them up and you have simply U, standing for *unworldliness.* It is a measure of anything and everything that doesn't fit into the scheme of physics. The master equation says the unworldliness is 0. You can unpack all of physics from this.

$U = 0$ is simpler (more "beautiful") than any other equation. It says everything the other equations do, and it's as simple as an equation can possibly get. An equation means you've got an equals sign, with one thing on the left of it, and one thing on the right. Three characters is the bare minimum, and $U = 0$ delivers that beautiful (anorexic?) limit.

Feynman's real point was that "$U = 0$" is a silly kludge serving no purpose except to say everything about the universe as concisely as possible! Feynman was asking, are you *sure* that's what you mean by beautiful? It's worth thinking about. A good answer to this interview question might start with Feynman's $U = 0$. Then, if you think you've got a better notion of what "beauty" is, describe it and tell what equation best fits it.

Chapter Five

? You want to make sure that Bob has your phone number. You can't ask him directly. Instead you have to write a message to him on a card and hand it to Eve, who will act as go-between. Eve will give the card to Bob, and he will hand his message to Eve, who will hand it to you. You don't want Eve to learn your phone number. What do you ask Bob?

Even if you give the short, simple answer (see pages 71–72), you may be asked to supply the RSA answer, too. It's not that complicated as long as Bob has a computer and can follow directions. You might ask the interviewer about Bob's maths and computer skills.

With RSA, each person generates two keys, a public one and a private one. A public key is like an e-mail address. It allows anyone to send you a message. A private key is like your e-mail password. You need it to get your e-mail messages, and you must keep it secret—otherwise, anyone could read your mail.

You won't be able to send Bob a secret message because he hasn't set up his keys. He may not know what RSA is until you tell him! But you don't need to send Bob a secret message. You want Bob to send *you* a secret message, namely your phone number. This means that you need keys for yourself, not for Bob. The outline of the solution is this:

> Hey, Bob! We're going to use RSA cryptography. You may not know what that is, but I'll explain exactly what you have to do. Here is my public key.... Take this and my phone number and produce an encrypted number by following these directions.... Send that encrypted number back to me, via Eve.

The trick is to word the instructions so that almost anyone can do it. You also have to be concise.

RSA cryptography was first described, it now appears, in 1973. Its original inventor was the British mathematician Clifford Cocks, who worked for Her Majesty's secret service. His scheme was considered impractical: it required a computer, of all things. That was not easy to come by when spies generally had to make do with cameras hidden in cuff links. Cocks's idea was classified until 1997. Meanwhile, in 1978 three MIT computer scientists independently came up with the same idea. The last initials of the MIT group—Ronald Rivest, Adi Shamir, and Leonard Adelman—supplied the acronym.

In the RSA system, a person who wants to receive messages must pick two random prime numbers, p and q. The numbers must be large and at least as big (in digits) as the numbers or messages being transmitted. For a phone number of ten decimal digits, p and q each should be at least ten digits.

One way to choose p and q is to Google a website that lists large prime numbers. The Primes Pages, run by Chris Caldwell of the University of Tennessee at Martin, works well. Pick two ten-digit primes at random. Here are two:

1,500,450,271 and 3,367,900,313

Call these p and q. You have to multiply them and get the exact answer. This is a little tricky. You can't use Excel or Google calculator or most other consumer software because they show a limited number of significant figures. One option is to multiply by hand. An easier one is to use Wolfram Alpha (www.wolframalpha .com): Just type in

1500450271 * 3367900313

and it will give the exact answer,

5053366937341834823

Call this product N. It's one component of your public key. The other component is a number called e, an arbitrarily chosen number, ideally of length equal to N, that does not divide evenly by $(p - 1)(q - 1)$. I may have lost you with that last part, but don't worry. In many applications, coders simply pick 3 for e. This is good enough for most purposes, and it permits fast enciphering.

Having chosen N and e, you're good to go. You just need to send those two numbers to Bob, along with the Complete Idiot's Guide to RSA Cryptography. Bob has to compute

$$x^e \bmod N$$

where x is the phone number. Since we've chosen 3 for e, the part on the left is x cubed. It will be a thirty-digit number. The "mod" indicates modulo division, meaning that you divide x^3 by N and take only the remainder. This remainder must be in the range of 0 to $N - 1$. Thus it's probably going to be a twenty-digit number. This twenty-digit number is the encrypted message that Bob sends back to you.

Bob therefore needs to be able to cube a number and do long division. The crucial part of the instructions could say

> Bob, I need you to follow these instructions carefully without questioning them. Pretend my phone number is a regular ten-digit number. First, I need you to cube the number (multiply it by itself and then multiply the product by the original number again). The answer, which will be a thirty-digit number, has to be exact. Do it by hand if you have to, and double-check it. Then I need you to do the longest long division you've ever done. Divide the result by this number: 5,053,366,937,341,834,823. The division also has to be exact. Send me the remainder of the division only. It's important that you don't send the whole part of the quotient—just the remainder.

Assuming Bob has Internet access (a pretty safe assumption, right?), the message could say

> Bob, go to this website: www.wolframalpha.com. You'll see a long, rectangular box outlined in orange. Type my ten-digit phone number into the box, without using dashes, dots, or parentheses—just the ten digits. Immediately after the phone number, type this:
>
> ^3 mod 5053366937341834823
>
> Then click the little equals sign in the right of the box. The answer, probably a twenty-digit number, will appear in a box labeled "Result." Send that answer, and only that answer, to me.

Naturally, Eve reads these instructions, and she also reads Bob's reply. She can't do anything with it. She's got a twenty-digit number that she knows is the remainder when the cube of the phone number is divided by 5,053,366,937,341,834,823. Nobody has yet figured out an efficient way to recover the phone number.

How are you any better off? You are because you have the *secret decoder key.* This, d, is the inverse of e mod $(p - 1)(q - 1)$. There is an efficient algorithm for calculating that—provided, of course, that you know the two primes, p and q, that were used to generate N. (You know them because you picked them, remember?)

Call Y the encoded number/message Bob sends back. His original message is

$$Y^d \bmod N$$

To figure this, you just type it into Wolfram Alpha (replacing Y, d, and N with the actual numbers).

Eve knows N, since it was on the card you asked her to give to Bob. She knows y, since it was Bob's reply to you. But she doesn't know d, and she has no way of learning it. Eve has algorithm trouble. It is easy to multiply two numbers—heck, they teach that to schoolkids. It is hard to factor a large number.

? If you had a stack of pennies as tall as the Empire State Building, could you fit them all in one room?

This may sucker you into thinking that it is one of those interview questions where you're intended to estimate an absurd quantity. Hold on—the question doesn't ask *how many* pennies. It asks, *will the stack fit in a room?* The interviewer wants a yes-or-no answer (with explanation, of course).

That should be a clue, as should the fact that the question doesn't say how big the room is. Rooms come in all sizes. Intuition might suggest that the stack wouldn't fit in a phone booth but would fit easily in the Hall of Mirrors at Versailles.

The answer is roughly this: "The Empire State Building is about a hundred stories tall [it's 102 exactly]. That's at least a hundred times taller than an ordinary room, measured from the inside. I'd have to break the skyscraper-high column of pennies into about a hundred floor-to-ceiling-high columns. The question then becomes, can I fit about a hundred floor-to-ceiling penny columns in a room? Easily! That's only a ten-by-ten array of penny columns. As long as there's space to set a hundred pennies flat on the floor, there's room. The tiniest New York apartment, an old-style phone booth, has room."

Swagger counts. The goal is not just to get the right answer but to make it look easy. Great athletes do this naturally. Lately, job seekers are expected to do the same.

? You have ten thousand Apache servers, and one day to generate £1 million. What do you do?

The Microsoft answer: Take the opportunity to regale the interviewer with your pet unrealized business plan. Expect the interviewer to listen politely and then ask, "Yes, but are you sure you can make a million pounds *the first day?*"

Note: Google is a server-intensive business that took about five years to turn a profit. YouTube *might* be turning its first real profits by the time you read this.

A relatively believable business plan would be high-speed security trading. It's reported that smart operators make millions every trading day by buying and selling securities that they hold, on average, for seconds. They normally liquidate all their holdings just before the markets close, so you'd have your profit by the end of the day. Such a scheme requires software (capable of out-arbitraging all the other trader bots out there) and fast hardware but wouldn't need anywhere near ten thousand servers.

The Google answer: Sell the servers for at least £100 each. That will "generate" £1 million, or more like £10 million. Then, if you've got some great business plan, use that as seed money. It will keep you going long enough to interest a venture capitalist (who's smart enough to know that great ideas don't make a million the first day).

? There are two rabbits, Speedy and Sluggo. When they run a 100-meter race, Speedy crosses the finish line while Sluggo is at the 90-meter mark. (Both rabbits run at a constant speed.) Now we match them up in a handicapped race. Speedy has to start from 10 meters behind the start (and run 110 meters), while Sluggo starts at the usual mark and runs 100 meters. Who will win?

The Microsoft approach: Let Speedy's speed be x and Sluggo's speed $0.9x$…

The Google approach: Speedy covers 100 meters in the time that Sluggo covers 90. In the handicapped race, Speedy starts from −10 meters. His 100 meters therefore take him to the 90-meter post. Meanwhile, Sluggo will have gone 90 meters, and since he started from 0, that will also put him at the 90-meter point. At this moment, the two will be tied. It's as if they're starting a new race at the 90-meter post, using 100 meters as the finish line. Naturally, the faster rabbit wins. That's Speedy.

? You've got an analog watch with a second hand. How many times a day do all three of the watch's hands overlap?

This is an update to a classic Microsoft interview question that asks how many times a day a clock's hour and minute hands overlap. Because that one's become pretty well known, interviewers have started using this variant.

The Microsoft answer: First, figure when the hour and minute hands overlap. Everyone knows the minute and hour hands overlap at 12:00 midnight and at approximately 1:05, 2:10, 3:15, and so forth. There's an overlap in every hour except 11:00 to 12:00. At 11:00, the faster minute hand is at 12 and the slower hour hand is on 11. They won't meet until 12:00 noon—ergo, there is no overlap in the 11:00 hour.

There are thus eleven overlaps in each 12-hour period. They are evenly spaced in time (since both hands travel at constant speeds). That means that the interval between overlaps of hour and minute hands is 12/11 of an hour. This comes to 1 hour, 5 minutes, and 27 3/11 seconds. The eleven alignments of minute and hour hands in each 12-hour cycle take place at

12:00:00

1:05:27 3/11

2:10:54 6/11

3:16:21 9/11

4:21:49 1/11

5:27:14 4/11

6:32:43 7/11

7:38:10 10/11

8:43:38 2/11

9:49:05 5/11

10:54:32 8/11

How can we determine if any of these times is a three-way overlap? Though the question is about an analog clock, think of a digital clock that gives time in hours, minutes, and seconds:

12:00:00

There is an overlap between minute and second hands only when the minutes figure (00 here) equals the seconds figure (00). There is a precise three-way overlap at 12:00:00. In general, the minute- and second-hand overlap will occur at a fractional second. For example, here,

12:37:37

the second hand would be at 37 past the minute, while the minute hand would be between 37 and 38 past the hour. The instant of overlap would come a split-second later. But the hour hand wouldn't be near the others, so this isn't a three-way overlap.

None of the hour- and minute-hand overlaps in the list above passes this test except for 12:00:00. That means all three hands align just twice every day, at midnight and noon.

The Google answer: The second hand is intended for timing short intervals, not for telling time with split-second accuracy. It's normally not in sync with the other two hands. "In sync" would mean that all three hands point to 12 at the stroke of midnight and noon. Most analog watches and clocks do not let you set the second hand from the stem. (I've never seen one that does.) A work-around would be to take the battery out (or let a windup watch run down), set the minute and hour hands in sync with where the second hand stopped, and wait until it becomes the time shown to replace the battery or wind up the watch. It would take a maniacal analog-watch fetishist to do that. But unless you do this, the second hand will not show the "real" time. It will be offset from the accurate seconds by a random interval of up to 60 seconds. Given a random offset, the odds are overwhelming that the three hands would *never* align precisely.

? You're playing football on a desert island and want to toss a coin to decide the advantage. Unfortunately, the only coin on the island is bent and is seriously biased. How can you use the biased coin to make a fair decision?

The Microsoft answer: Toss the coin a great number of times to determine the percentage of heads and tails. (Insert discussion of statistical significance here.) Once you know that the coin comes up heads 54.7 percent of the time (with error bars), you use that fact to design a multiple-toss bet with odds as close to even as desired. It will be something like, "We toss the coin a hundred times and heads has to come up at least fifty-five times for team A to win the advantage; otherwise team B gets it."

The Google answer: Toss the coin twice. There are four possible outcomes: HH, HT, TH, and TT. Since the coin favors one side, the chance of HH will not equal the chance of TT. But HT and TH must be equally probable, no matter what the bias. So toss twice, after agreeing that HT means one team gets the advantage and TH means the other does. Should it come up HH or TT, ignore it and toss another two times. Repeat as necessary until you get HT or TH.

Besides being simpler, this scheme is incontestably fair. The Microsoft scheme only approximates fifty-fifty odds.

Chapter Six

? How much would you charge to wash all the windows in Seattle? (Fermi question.)

The first step here is to estimate the population of Seattle. The U.S. Census recently put it at 594,000 (city limits) or 3.26 million (metro area). In a job interview, no one would fault you for saying Seattle has about a million people.

How many windows are there per Seattle resident? In Manhattan, young people count themselves lucky to have one window. Seattle is different; apartments are larger, and more people live in houses with panoramic windows overlooking evergreen forests. Many houses and townhouses are two story. A decent guess, favoring the convenient round number, is ten residential windows per Seattleite.

There are also windows at places of work, Starbucks, department stores, airports, concert halls, and so forth. This probably doesn't add all that much to the per capita total. The average cubicle has no windows. A big-box store has little surface area (and few windows) relative to its volume. The windows in public spaces like restaurants and airports are shared among the huge mass of people using them.

Don't forget windows in cars. (You might ask the interviewer whether to count them.) A car is going to have four windows at bare minimum, often twice that. But big 4x4s are driven by big families and don't add many windows per capita.

A reasonable guess is that windows outside the home account for another ten per person. This comes to twenty windows per Seattle resident. Assuming a population of a million, there are about twenty million windows to be cleaned.

How much should you charge to wash a window? With the windows in your home, it takes a few spritzes of Windex, a few paper towels, and a few seconds. Some of the windows in Seattle are huge, like those in the restaurant atop the Space Needle, and they're high up, requiring special crews, special equipment, high workers' comp rates, and considerable guts.

Someone knowing what he's doing could probably clean one side of a typical window a minute, given that most are small. That means cleaning "a window" (both sides) in two minutes. That comes to thirty windows an hour.

Say the average window washer makes $10 an hour. Throw in

another $5 an hour for supplies and insurance. That's $15 for an hour's work, which cleans thirty windows. Cost per window: 50 cents.

Twenty million windows times 50 cents is $10 million.

This question is used at Amazon and Google. Just so you don't miss the joke, such as it is, Windows is another company's registered trademark.

? A man pushed his car to a hotel and lost his fortune. What happened? (Lateral thinking puzzle.)

He was playing Monopoly.

? You get on a ski lift at the bottom of the mountain and take it all the way up to the top. What fraction of the lift's chairs do you pass? (Classic logic puzzle.)

You pass *all* the ski lift's chairs (except, of course, your own). The lift is like a loop of string on a pulley. The chairs are suspended from all parts of the loop. Because one half of the loop carries chairs downward as the other half carries them upward, chairs pass you with a relative speed that is twice that of the pulley itself. In taking the lift up, you traverse just half of the complete loop. But since the relative speed is twice the pulley's speed, you pass 100 percent of the loop and thus all the chairs other than your own.

You might wonder how you can pass the chair just in front of you. Moments before you reach the top of the mountain, the chair in front transfers to the return half of the loop. It's then going downward, and it passes your chair, going upward, just before you get off.

? Explain what a database is to your eight-year-old nephew, using three sentences. (Test of divergent thinking.)

A database makes it convenient to find information (not just store it, which is easy in comparison). The trick is to think of creative analogies relevant to an eight-year-old. You probably start by brainstorming: "A database is like...a Rolodex (does anyone under fifty have one?)...a magic wizard of information (too patronizing?)...an iPod...a TiVo...." Pick the best analogy and work it into a three-sentence answer:

> A database is an iPod for information. With an iPod, you can store thousands of songs and still find any track you want quickly. A database does the same thing with information that people have stored on a computer or the Internet.

? Look at this sequence:

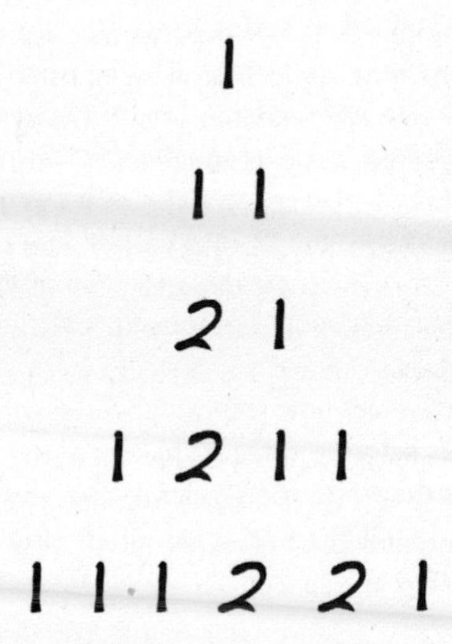

What's the next line? (Insight question.)

The interviewer writes this on the whiteboard. There's just enough pattern to drive maths nerds nuts. Hint: try reading the lines out loud.

This is the *look-and-say sequence*, described by the mathematician John Horton Conway in 1986. Except for the first line, each inventories the line above it. The third line, for instance, could be read as "two one[s]." Look at the line above it. It's two 1s.

The lowermost line you're given consists of three 1s, two 2s, and one 1. The following line must be

3 1 2 2 1 1

This puzzle appeared in the *Google Labs Aptitude Test,* a mock test distributed to university students as a recruitment promotion in Autumn 2004. HR departments usually discourage insight questions, but some interviewers can't resist. For the record, the look-and-say sequence is not just a one-shot maths joke. Conway proved some original and (semi)serious results about it. Engineers will recognize the series as a form of run-length encoding. When compressing video for an episode of *South Park,* you don't store the color of every pixel of Kyle's green hat. Instead you use run-time encoding, which basically says something like "the next 452 pixels are all the same shade of green."

? You have twenty-five horses. How many races do you need to find the fastest three horses? You don't have a timer, and you can run only five horses per race. (Algorithm question; could also be called a logic puzzle.)

You might begin by asking the interviewer whether you can assume the "fastest" horse always wins a given race. It doesn't work that way at the racetrack. But it simplifies this puzzle greatly to assume that when A beats B in one race, A is objectively and indisputably faster. You'll be told that it's okay to make that assumption, that the race is indeed to the swiftest.

The usual first thought is that you'll need at least five races. Any of the horses could be in the top three. Therefore you have to race all twenty-five. There's no way of doing that in less than five races of five horses each.

Fair enough. Second conclusion: five races isn't enough. Divide the twenty-five horses into groups of five and race them, each horse competing once, against four others. One of the races might come out like this:

1. Seabiscuit
2. Northern Dancer
3. Kelso
4. War Admiral
5. Dancer's Image

You can't conclude that Seabiscuit is the fastest horse of the whole twenty-five, or that he's even in the top three. To give an extreme counterexample, it's conceivable that the slowest horses in the other races are *all* faster than Seabiscuit (who might rank as low as twenty-one out of twenty-five overall).

Have we learned anything from this race? Of course we have. We've learned how these particular five horses rank. We've also learned that we can rule out War Admiral and Dancer's Image. Since they didn't make the top three in this race, they can't be in the top three out of the entire field of twenty-five.

The same goes for the fourth- and fifth-place horses in the other races. Every five-way match rules out two horses as contenders for the fastest three. After the first set of five races, we can scratch ten horses, leaving fifteen in the running as candidates for the fastest three.

The sixth race will have to test horses who have performed well in the five initial heats. A reasonable plan is to race the five number one horses against one another. Let's do it. Take Seabiscuit from the race above and pit him against the winners of the other races. The outcome might look like this:

1. Easy Goer
2. Seabiscuit
3. Exterminator
4. Red Rum
5. Phar Lap

Once again, we can rule out two horses, Red Rum and Phar Lap. They cannot possibly be in the fastest three out of twenty-five, given this outcome. We also learn that Easy Goer is the fastest horse of all, fastest of the number one fastest horses! Had the question simply asked for fastest horse out of twenty-five, the answer would be Easy Goer.

We need the three fastest, though. Not only can we rule out Red Rum and Phar Lap, but we can rule out *all* the horses they beat in the first races they ran. The horses they beat are slower, and we already know Red Rum and Phar Lap won't make the grade.

Look at Exterminator. Since he came in third in this race, any horses he beat in his first race are out of the running, too.

Then there's Seabiscuit. Based on this latest race, he might be, at best, the second-fastest horse of all. That leaves open the possibility that Northern Dancer, who placed after Seabiscuit in the first round, could be the number three horse overall. (The ranking would then be Easy Goer, Seabiscuit, Northern Dancer.) Kelso, who came in third in Seabiscuit's first race, is now out of the running.

The two horses that came in second and third after Easy

Goer in his first race are still contenders. It's still possible that those horses are faster than Seabiscuit, for they've never raced him.

In short, there are now six horses in the running. They're the top three in this race; the two who came in second and third to this race's number one winner in his first race; and the one who came in second to this race's number two horse in his first race.

We already know that Easy Goer is the fastest horse of all. For that very reason, there's no point in racing him again. That leaves just five horses. Naturally, we race those five in the seventh and final match. The top two horses in that seventh race are the second and third overall.

To recap: Start with a qualifying round of five races, in which all twenty-five horses compete once. Follow that with a championship race, limited to the winners of the qualifying-round races. The winner of this race will be the number one horse overall. Follow that with a final race of the remaining five horses who are still logically in the running. The horses that win and place second in this match will be the number two and number three horses overall.

Chapter Seven

? Imagine you have a rotating disk, like a CD. You are given two colors of paint, black and white. A sensor fixed to a point near the disk's margin can detect the color of paint underneath it and produce an output. How do you paint the disk so that you can tell the direction it's spinning just by looking at the sensor's readings?

The first thing to understand is, you can't look at the disk. You're in London, the disk is on Mars. You have to determine the spin direction based on the sensor telemetry and nothing more.

The sensor will report the color of the spot directly underneath it at successive moments. The readout will be something like *black... black... black... white... white....* The goal is to paint the disk so that the readout is *not* the same forward and backward. To put it another way, the readout must not be a palindrome.

A palindrome is one of those funny words or phrases that reads the same forward and backward. Examples are *mum, racecar, step on no pets,* and *sit on a potato pan, Otis.* It usually takes effort to think up a palindrome, whereas *not* speaking in palindromes is as easy as falling off a log! You might therefore think it would be easy to devise a paint pattern that's not a palindrome. But there are two complications. The question gives you only two "letters" to work with, B and W (for black and white paint). Second, you must avoid a *circular* palindrome as well as the regular kind.

For example, you wouldn't want to paint the disk half-black and half-white. Then the readout would be something like *black... black... black... white... white... white... black... black... black... white... white... white....* This is not a regular palindrome, but it is a circular palindrome in that if you join one end to the other, you get the same result when reading clockwise or anticlockwise. Looking at just the endless data stream, there's no way of telling which way a half-black, half-white disk is spinning.

Not all patterns are circular palindromes. Given three colors, the disk could be painted in three equal slices of black, white, and red (in that clockwise order). Then a clockwise spin would produce *black... black... black... red... red... red... white... white... white....* An anticlockwise spin would give *black... black... black... white... white... white... red... red... red....* These are easily distinguished. In the first, a sequence of red readings immediately follows black readings; in the second, red follows white. (It's something like the rule hikers use to distinguish the deadly coral snake: "Red touch yellow, kill a fellow.")

The question doesn't permit red paint, but you can manage

zebra stripes. Paint one sector with many thin slices of alternating black and white. Then you can detect whether the stripes come immediately after the black sector (clockwise spin) or after the white sector (anticlockwise).

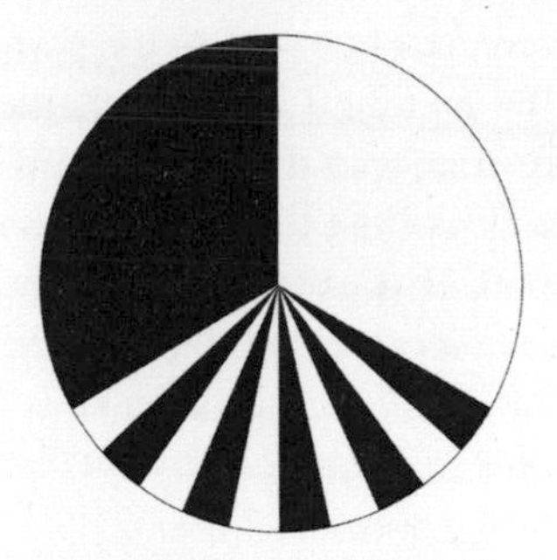

Good: Paint one-third of disk with "zebra stripes"

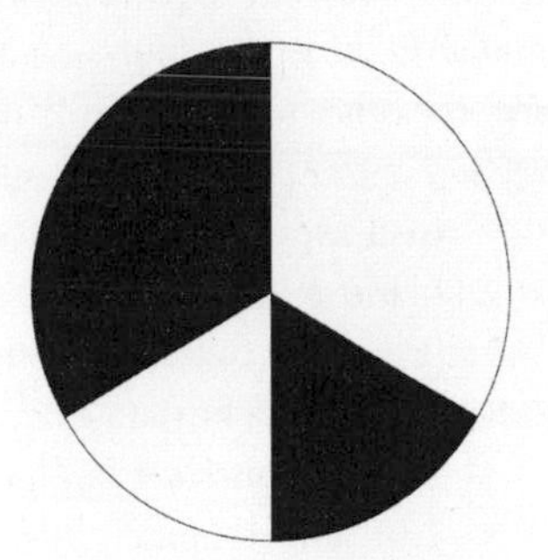

Better: Paint two unequal sectors of each color

This answer can be improved. The interviewer hasn't said how fast the disk will be spinning or how promptly the sensor can register a change of color (its "shutter speed" or "exposure delay"). The disk might be spinning so fast that the sensor records the color of one stripe flitting under it and skips the others. This could create misleading readings in the striped zone.

It's good design to keep the stripes as few and broad as possible. In fact, two stripes in the striped zone will suffice. Each will be a 1/6 slice of the whole disk, and you paint them to contrast with their immediate neighbors, of course.

When the disk is colored the way shown here, and when the sensor is able to take six readings per revolution, a clockwise spin will look like *black... black... white... black... white... white...* and anticlockwise will be the reverse.

There's a variation of this question where the disk is already painted, half-white and half-black. You have an unlimited number of sensors. How many sensors do you need to place around the disk to tell the spin direction?

All you can tell from one sensor is the percentage of black and white, 50-50—which we already know. With two sensors, the first impulse might be to place them at opposite sides of the disk, at 12:00 and 6:00. At any given moment, whatever color one sensor is reporting, the other will necessarily report the opposite. Consequently, the second sensor conveys no useful information with this arrangement.

Instead, you should place the sensors close to each other, say at 2:00 and 2:01. Most of the time, both sensors will see the same color. However, one will get the news of a color change before the other one does. The data stream will look something like this:

2:00 sensor: *black... black... white... white... white...*

2:01 sensor: *black... black... black... white... white...*

This indicates that the black-white boundary hits 2:00 before it does 2:01. Therefore the boundary, and the disk, must be spinning clockwise. Should the 2:01 sensor see the change first, the disk is spinning anticlockwise.

? How many lines can be drawn in a plane such that they are equidistant from three noncollinear points?

The interviewer is yanking your chain with the language. The vocabulary word you'll need is *noncollinear.* Pick up a marker and make three dots on the whiteboard. Just make sure they're not in a straight line (which is all that *noncollinear* means).

There are your three noncollinear points.

The question asks, how many lines can you draw that are equidistant from those three points? *Equidistant* means "the same distance from." This is another common stumbling block. A line zooms off to infinity in both directions. So how can a line be "the same distance from" a point? The only interpretation that makes sense is that we are talking of the minimum distance between the line and the point. It's like asking how far a summer rental is from the beach. You measure the shortest straight-line distance from house to surf.

Here's an example. I've drawn a line between the three points so that the minimum distance between each point and the line (shown dotted) is the same.

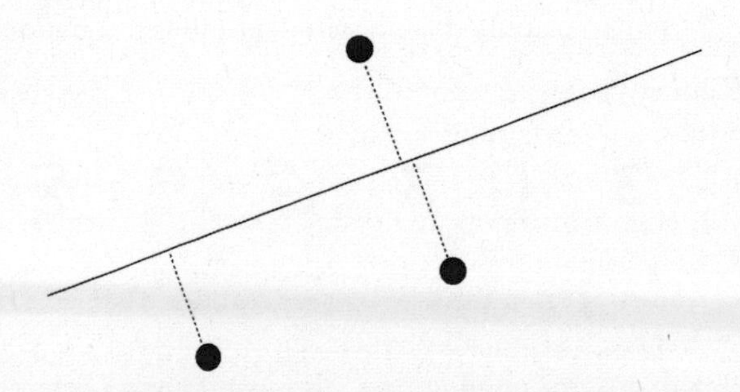

In order to draw this line, I made a simple geometric construction (well, actually, I eyeballed it, as you will on the whiteboard). First imagine a line between the two lowermost points. Then draw a line parallel to that, halfway between the imaginary line and the point at the top.

You can repeat this construction two more times, drawing an imaginary line between each pair of points. (See the following page.) They look like this. The answer to the question is that there are three equidistant lines.

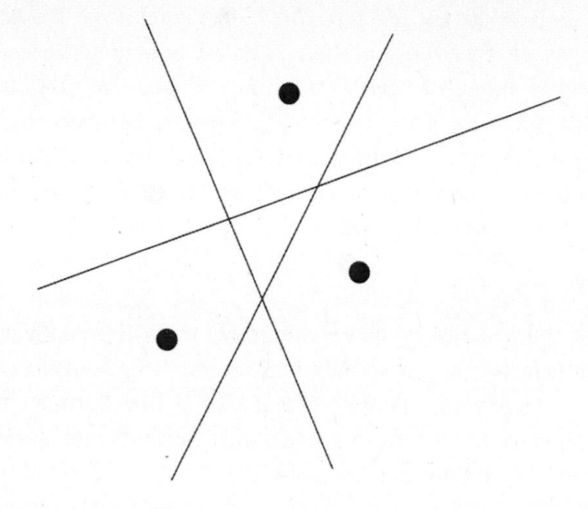

? Add any standard arithmetic signs to this equation to make it true.

$$3 \quad 1 \quad 3 \quad 6 \quad = \quad 8$$

Starting from the left, you see 3 and 1. The first maths sign you learned was probably a plus sign. 3 + 1 gives 4. Conveniently, 4 is half of 8, the number that's the goal. Moving right, there's a 3 and a 6. Well, 3 is half of 6. Put a division sign between them and you've got 3/6 or 1/2. *Dividing* by 1/2 is the same as multiplying by 2. This gives the answer

$$(3+1) \div (3 \div 6) = 8$$

That wasn't so hard, was it?

Now for the real test. The hidden agenda of this question is to see whether you stop there or go on. Some interviewers will be helpful enough to give you a hint ("That's one good answer.... Can you think of any others?") So copy the numbers elsewhere on the whiteboard and start anew. The more distinct and clever answers you can come up with, the better (until the interviewer stops you).

For instance, there's

$$((3+1)\div 3)\times 6 = 8$$

Many miss this one because of what might be called anti-fraction bias. They see the expression in parentheses comes to 4/3 and give up, figuring they won't be able to get the whole number 8 out of it.

There's also a creative answer using the keyboard's exponentiation operator (^).

$$(3-1)^\wedge(-3+6) = 8$$

Square root signs add other possibilities. There's

$$3-1+\sqrt{36} = 8$$

and

$$\sqrt{3-1\div(3\div 6)} = \sqrt{8}$$

Then there is a legion of semifacetious approaches like adding a greater-than sign to change the equals sign to a greater-than-or-equal-to sign.

$$3 + 1 + 3 + 6 >= 8$$

Can you think of any other good answers?

? There's a bar where all the customers are antisocial. It has 25 seats in a line. Whenever a customer comes in, he invariably sits as far from the other customers as possible. Nobody will sit directly next to anyone else—should a customer come in and find that there are no such seats, he walks right out. The bartender naturally wants to seat as many customers as possible. If he is allowed to tell the first customer where to sit, where should that be?

The densest-possible packing is to alternate customers and empty seats, with customers on both ends of the bar. This would put someone in all the odd-numbered seats, including the end seats #1 and #25, and leave all the even-numbered seats empty. That's thirteen customers.

This arrangement doesn't just happen. Suppose the first customer plunks down in seat #1. The next antisocial barfly will pick #25, since that's as far as he can get from #1. The third customer will have to go in the exact middle, seat #13. The two after that will fill in the gaps, seating themselves in #7 and #19. So far, so good.

Eventually someone will want to sit between the customers already in seats #1 and #7. He'll pick seat #4 because that puts 2 empty seats between him and his nearest neighbors. But no patrons will sit around him. The rest of the bar will fill in the same way, with 2-seat gaps between each customer, making this the *least* efficient scheme possible (seating nine rather than the optimum thirteen).

Many puzzles, including this one, are best solved by working backward. We know the seating plan we want to end up with; how do we get there?

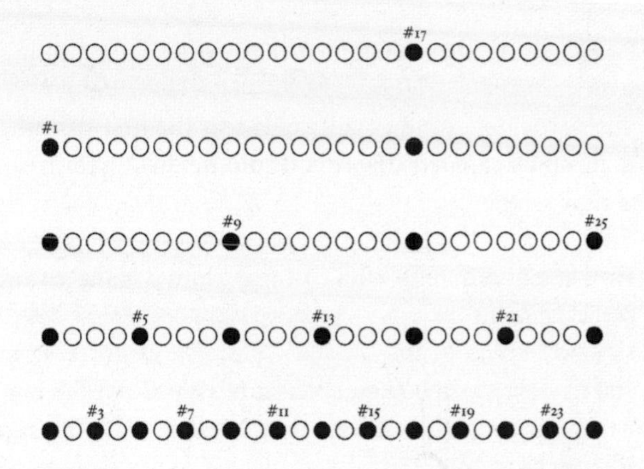

As the diagram shows, there's a lot of symmetry, as in the growth of a crystal. Little subsections of the bar fill in in the same way. Focus on the low-numbered end of the bar. We need to have people in seats #1 and #5 because that will allow someone to sit in #3.

How do you make sure someone will sit in seat #5? Answer: Have customers in seats #1 and #9. Five will then be the middle seat, maximally distant from #1 and #9.

How do you get someone in #9? Have customers in #1 and #17. And how do you get someone in #17? Well, the bar isn't long enough to force the choice by having customers in #1 and #33. So the bartender has to tell the first customer to sit in seat #17. That's the answer to the puzzle.

Here's how it plays out. The first guy sits in #17 (top row of diagram). The second sits as far away as possible, in seat #1.

The third customer has two choices: seat #9 or #25. Both are 7 empty seats away from another customer. It would be in the sullen, standoffish spirit of the place for him to favor #25, with one distant neighbor rather than two, leaving #9 for the fourth customer.

The next three customers split the difference between the first four, taking seats #5, #13, and #21. Each is 3 empty seats away from a neighbor.

Finally, the next six customers occupy the 6 remaining seats having no immediate neighbor: #3, #7, #11, #15, #19, and #23.

The bartender could equally well seat the first customer in seat #9, in which case the diagram would be the mirror image of the one shown here.

? How many different ways can you paint a cube with three colors of paint?

Start by pretending the cube is a Nissan Cube—a chunky car for the Generation Y market—and Nissan is offering custom paint jobs on its website. Buyers can specify any of three designer colors for each of the car's six sides. That's front, back, right, left, roof, and, yes, even the underside. Since there are three color options for each side, there are $3 \times 3 \times 3 \times 3 \times 3 \times 3$ possible colorings overall. That's 729.

This straightforward calculation follows from the fact that all six sides of a Nissan Cube are distinguishable. The interview question concerns a more abstract cube, like a one-inch block you could hold in your palm. The cube's faces have no distinguishing features. Therefore the number of meaningfully distinct paint jobs is much less than 729. The 729 figure counts as different all six cases where one face is painted red and the others are white. It makes more sense to count that as one paint job. Otherwise, you're like a diner who complains to the waiter that his menu is upside down. Turn it over!

As long as one color configuration can be rotated into another, it should count as the same. This makes the question a lot harder. The way to tackle it is to divide and subdivide the problem into manageable chunks. No special maths ability is required, just persistence and organization. This question isn't about spotting geniuses. It's about finding out who gives up, who gets hopelessly confused, and who gets the job done. You'll probably use the whiteboard in two ways: to tally the number of paint schemes and to draw some diagrams.

Ready to begin? Take a deep breath. The number of ways of painting a cube with three colors equals

- the number of ways of painting a cube with exactly one color (chosen from the set of three), *plus*
- the number of ways of painting a cube with exactly two colors (chosen from the set of three), *plus*
- the number of ways of painting a cube with exactly three colors.

Obviously, there's only one way of painting a cube with a single color of paint. You paint every face the given color, and that's that. As we have three colors of paint to draw on, there are three monochrome color schemes: all-white, all-black, and all-red, let's say.

Now for the two-tone schemes. For the sake of concreteness, I'll start by picking black and white as the two colors. It doesn't take much mechanico-spatial imagination to list the possible black-and-white schemes:

- Paint a single face black (leaving the others white).
- Paint two faces black, and they're adjacent to each other, sharing an edge.
- Paint two faces black, and they're opposite each other.
- Paint three faces black so that they fit around a corner.
- Paint three faces black so that two are opposite each other. The remaining black face adjoins the two others. (If you could peel off the black faces, they'd form a three-by-one rectangle, like a comic strip.)

- Paint four faces black so that the remaining two white faces are adjacent to each other.
- Paint four faces black so that the two white faces are opposite each other.
- Paint five faces black, leaving one white.

That makes eight distinct ways of painting a cube in black and white. It follows that there must also be eight ways of painting it red and white, and eight ways of painting it black and red. That totals twenty-four ways to paint a cube in exactly two colors.

The real challenge is a cube painted with all three colors. At least you don't have to choose a color palette this time.

Subdivide further. Either the three colors are used equally, on two faces each — or they're used on three, two, and one face, respectively — or on four, one, and one face.

Start with the equal case of two faces per color. Here are the possibilities:

- Every face is opposite a face of the same color.
- Every face is adjacent to a face of the same color (and therefore, no face is opposite a same-color face).
- The white faces (only) are opposite each other. The two black faces, and the two red ones, are adjacent to each other.
- Only the black faces are opposite each other.
- Only the red faces are opposite each other.

There are two mirror-image versions of the second case, with each face adjacent to a same-color face. This is akin to the two mirror-image ways of numbering the faces of a die. The two versions can't be rotated into each other, and that makes six two-two-two paint schemes in all. Now for the three-two-one cases.

This is the toughest situation to visualize, so you'll probably want to draw it. Start by assuming there are three white faces, two black faces, and one red face. Here are the possible schemes:

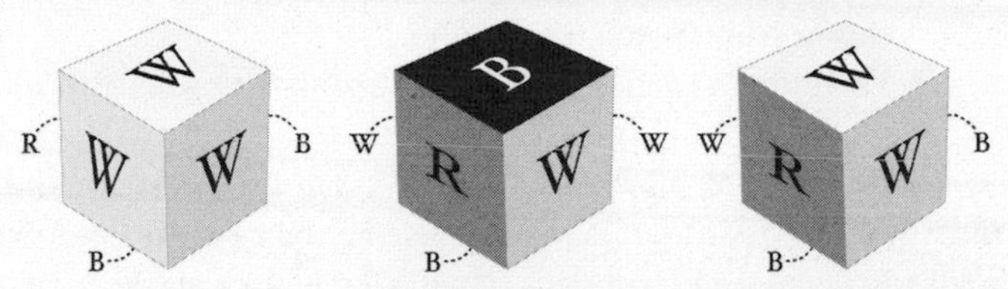

- The three white faces meet at a corner. No matter which of the remaining faces you choose to paint red, you can rotate the cube to be identical with the other choices. Therefore, this counts as one paint scheme.
- Two of the white faces are opposite each other, and the red face is opposite the remaining "middle" white face. The black faces are opposite each other.
- Two of the white faces are opposite each other, and the single red face adjoins all three white faces. This means that the two black faces adjoin each other. (The red face may go on the right or the left of the whites. Either way, you can rotate the cube to get the opposite.)

This makes three distinct possibilities for three white, two black, and one red. However, you can choose *any* of the three colors to be the one with three faces. Then you can pick either of the remaining colors to have two faces. This leaves the remaining color for the single face. Therefore, there are really $3 \times 2 \times 1 = 6$ color palettes for a three-two-one design. That makes $6 \times 3 = 18$ paint jobs overall. Finally, take the four-one-one cases. Say the cube is mostly white, with a single black and single red face. The latter can be opposite each other or adjacent. That's two possibilities. Ditto if the predominant color is black or red. That means there are $2 \times 3 = 6$ paintings. Add that to the six two-two-two color schemes and the 18 three-two-one schemes for a total of 30 three-color schemes.

Summing up (yes, I know—you really want this job, right?):

- There are 3 ways of painting a cube with exactly one color, chosen from three.
- There are 24 ways of painting a cube with exactly two colors out of three.
- There are 30 ways of painting a cube with exactly three colors.

The grand total, 57, is your answer.

Does it seem like today's job interviews have fallen behind the looking glass? A version of this puzzle was invented by Lewis Carroll in the late nineteenth century. Carroll asked how many ways there were to paint a cube with six colors of paint. The answer to that one is 2,226. This one is *much* easier.

Chapter Eight

? How many ridges are there on the rim of a U.S. quarter?

A quarter is about an inch in diameter. Its circumference is pi (3.14159+) times that. Say, three inches. The only uncertain part of the calculation is estimating how many ridges there are to an inch. It's got to be more than 10 and is probably less than 100. Take 50 as a reasonable value and multiply that by 3 to get the answer: 150 ridges.

The actual number on a quarter is 119, and by the way, the ridges are properly called reeds. They were originally put on gold coins to discourage grifters from shaving precious metal off the edges. It may be fitting that this question is asked at the Big Four accounting firm Deloitte.

? How many bottles of shampoo are produced in the world in a year?

People in affluent countries run through several bottles of shampoo a year. Many in developing nations can't afford such a luxury as shampoo. You might as well guess that it averages out to one bottle per person (unless you're interviewing at Procter and Gamble, the interviewer won't know any better). The answer is, there are about as many bottles produced per year as there are people in the world, 6 billion.

A handy pointer: It's hard to go too far wrong when estimating consumption of popular consumer products. Use your own consumption as a guideline and adjust. The resulting guesstimate won't be orders of magnitude off, and that's all that matters.

? How much toilet paper would it take to cover Greater London?

A square of toilet paper is about 4 × 4 inches. Nine squares, in a 3 × 3 grid, would then make a square foot. Let's call that "about 10" squares to a square foot. A roll of toilet paper has maybe 300 squares. Then a roll is about 30 square feet.

A mile is about 5,000 feet. A square mile is therefore 5,000 × 5,000 or 25 million square feet. The number of rolls of toilet paper needed to cover a square mile would be 25 million divided by 30. For Fermi question purposes, 25 is practically the same as 30. Call it a million rolls of toilet paper to the square mile.

Greater London is more or less the area contained by the M25 motorway, which is almost circular. It takes about two hours to go round it at 50mph. This makes its circumference 100 miles. As a circle's diameter is its circumference divided by pi (roughly 3), the M25's diameter is 100 ÷ 3, which is about 30 miles. So, its radius is 15 miles. A circle's area is given by πr^2, which means the area contained is approximately $3 \times 15^2 = 675$ square miles. To cover it, you'd need 675 × 1 million rolls. That's 675 million rolls.

? What is 2^{64}?

This is Google's harder version of a common tech-firm screener, "What is the value of 2^{10}?" Any engineer ought to know the answer to that (1,024). But nobody is expected to know 2^{64} offhand. It qualifies as a Fermi question because you're intended to do the maths in your head, and the answer doesn't have to be exact.

All right, 2^{10} is about 1,000. Multiply 2^{10} by itself six times and you have 2^{60}. That would be approximately 1,000 to the sixth power, or 10^{18}, also known as a quintillion (in the United States). You just have to multiply that by 2^4 to get 2^{64}. Well, 2^4 is $2 \times 2 \times 2 \times 2 = 16$. The quick-and-dirty answer is "about 16 quintillion."

It's a little bigger than 16 quintillion because 1,024 is 2.4 percent bigger than 1,000 is. We used this approximation six times, so the total would be more than 12 percent bigger. That adds another 2 quintillion or so. Call it 18 quintillion.

That figure should be good enough for any job outside of a calculating-prodigy variety act. The exact value is 18,446,744,073,709,551,616.

? How many golf balls will fit in a Mini?

We might check that the question means the new Mini rather than the original 1960s style. It's shorter than a person, so it's maybe 4.5 feet high, but with the clearance off the floor, let's say 4 feet. There's space for two people to sit side by side, with roughly 2.5 feet per person. So, that's 5 feet wide. It's probably two people long, which is approximately 10 feet. However, the bonnet takes up about half of the length, so the usable length is 5 feet. The interior volume, then, is about $4 \times 5 \times 5$, which is 100 cubic feet.

A golf ball is somewhat bigger than an inch in diameter. Let's say that 10 golf balls line up to make a foot. A cubic lattice of

$10 \times 10 \times 10$ golf balls, an even thousand, would just about occupy a cubic foot. That gives a quick answer of about $100 \times 1{,}000 = 100{,}000$.

You'll notice that many Fermi questions involve spherical sports equipment filling a Mini, swimming pool, jet plane, or sports stadium. You can score a bonus by mentioning Kepler's conjecture. In the late 1500s, Sir Walter Raleigh asked the mathematician Thomas Harriot to devise the best way to stack cannonballs on navy ships. Harriot mentioned the problem to his friend, the astronomer Johannes Kepler. Kepler in turn supposed that the densest way of packing spheres was the one already used for cannonballs and fruit. Start with a flat layer of spheres in a hexagonal array; then put another layer on top of that, fitting each new sphere into the depressions between three spheres on the lower layer. In a large crate, this arrangement approaches a maximum density of about 74 percent. That's the space occupied by the cannonballs or oranges, as a fraction of total space. Kepler guessed this was the densest packing possible, but he was unable to supply a proof.

Kepler's conjecture, as it was called, remained a great unsolved problem for centuries. In 1900 it made David Hilbert's famous list of twenty-three unsolved problems in mathematics. A number of people have claimed to prove it, including the architect Buckminster Fuller, of geodesic dome fame. All such resolutions were quickly rejected as wrong until 1998, when Thomas Hales offered a complicated, computer-assisted proof showing that Kepler was right. Most believe his result will stand up, though the construction of a formal proof is currently ongoing. Hales estimated it would take another twenty years.

I assumed above that each golf ball effectively rests in an imaginary Lucite cube whose edges equal the ball's diameter. You stack those Lucite cubes like building blocks. This would mean that the balls occupy about 52 percent of the space ($\pi/6$, to be exact, as you can compute from the formula for the volume of a sphere, $4/3\pi r^3$.) Break out of the imaginary Lucite boxes, and you can pack far more balls in a volume. This is an empirical fact. Physicists have done experiments by pouring steel balls into big flasks and calculating the density. The resulting random packing occupies anywhere from 55 to 64 percent of the space. That's denser than a cubic lattice, though well under Kepler's maximum of about 74 percent. It's also a pretty big range. How you fill the container matters. When the spheres are added gradually and gently, like sand pouring through an hourglass, the density is at the low end of the range. When the container is shaken vigorously, the spheres settle into a denser packing of up to 64 percent.

Where does this leave us? Someone willing to painstakingly stack the golf balls in the cannonball pattern can pack about 42 percent more golf balls in the Mini than you'd estimate from a cubic lattice. That seems an absurd amount of labor, even for an absurd question. The reported density of stirred random packings is a more realistic goal. You might achieve that by pouring golf balls into the Mini and stirring with a stick to settle them. That would give a density of about 20 percent more than that of a cubic lattice. You might therefore increase your final estimate by 20 percent, from 100,000 to 120,000.

For the record, miniusa.com states that a Mini Cooper Hardtop is 145.6 inches long, 75.3 inches wide (including mirrors) and 55.4 inches high. The regulation diameter of a golf ball is 1.690 inches, give or take 0.005 inches.

Chapter Nine

? It's raining and you have to get to your car at the far end of the car park. Are you better off running or not, if the goal is to minimize how wet you get? What if you have an umbrella?

To answer this question, you must reconcile two conflicting trains of thought. The case for running is this: The longer you are out in the rain, the more drops fall on your head, and the wetter you get. Running shortens your exposure to the elements and thereby keeps you drier.

There's also a case for *not* running. In moving horizontally, you slam into raindrops that wouldn't have touched you had you been standing still. A person who runs in the rain for one minute gets wetter than a person who just stands in the rain for a minute.

That valid point is mostly beside the point. You have to get to your car, and there's nothing to be done about that. Imagine yourself zipping across the car park at infinite speed. Your senses are infinitely accelerated, too, so you don't slam into cars. From your point of view, external time has stopped. It's like the "bullet time" effect in a movie. All the raindrops hang motionless in the air. Not a drop will fall on your head or back or sides during the trip. But to get to the car, you've got to carve a tunnel through the rain. The front of your clothing will sop up every single raindrop hanging in the path from shelter to car.

When you travel at normal speed, you're fated to run into those same raindrops or, rather, their successors. At normal speed, you also have drops falling on your head. The number of raindrops you encounter will depend on the length of your horizontal path and also on the time it takes to travel that path. The length of the path is a given. The only thing you can control is the time it takes. To stay as dry as possible, you should run as fast as possible. Running makes you less wet—*provided you don't have an umbrella.*

Had you an umbrella as wide as a city block, and were you able to hold it, it wouldn't matter whether you sauntered or sprinted. You'd be dry as toast.

Most umbrellas are barely big enough to keep the user dry

when he or she is standing in gentle, vertical rain. In practice, you expect to get a little wet.

Umbrellas work by creating a rain shadow, a zone where there are no raindrops. In a vertical downpour, and with a circular umbrella, the rain shadow is a cylinder. When the rain is coming at an angle, the rain shadow becomes a skewed cylinder. However, as every seasoned umbrella user knows, it's best to point the umbrella in the direction of a driving rain. This makes the rain shadow a proper cylinder again, now pointed at an angle to the vertical.

The standing human body doesn't fit so well into a slanted cylinder. Were a hurricane driving the rain at you horizontally, you would have to hold the umbrella horizontally, and a three-foot-diameter umbrella would protect only about half your body. The rest would get soaked.

Wind is bad, and so is motion. The skilled umbrella wielder knows to tilt the umbrella forward, in the direction of motion, to get the maximum coverage. In fact, wind and motion are indistinguishable as far as optimal umbrella pointing goes. Running at ten miles an hour in windless, vertical rain demands the same tilt as standing still in a ten-mile-an-hour wind. Either way, the raindrops are coming at you at ten miles an hour, horizontally, in addition to their downward velocity.

In a vertical rain, you're best off walking slowly. The umbrella will not have to be tilted much, and your body should fit within the rain shadow. Ideally, you should walk no faster than the speed where the rain shadow just covers your feet. Then you'd stay dry.

Reality is messier than that. There are always going to be gusts of wind, spatter from pavement, and runoff from the umbrella itself. The rain hitting the top of the umbrella does not disappear; it slides off the umbrella and falls in a cylindrical sheet encircling the rain shadow. There is more rain in that runoff zone than there is elsewhere. That means that any part of your body that intersects the runoff zone gets wetter faster than it would had you not used an umbrella at all.

The advantage of slowness diminishes in high headwinds. The umbrella has to be pitched at such an angle that your lower body is out of the rain shadow. You'll get half-soaked no matter what you do.

All that reasoning boils down to the advice you may have heard from Mum: walk if you've got an umbrella; run if you don't.

? You have a glass jar of marbles and can determine the number of marbles in the jar at any time. You and your friend play this game: In each turn, a player draws 1 or 2 marbles from the jar. The player who draws the last marble wins the game. What's the best strategy? Can you predict who will win?

The number of marbles decreases with each move and must ultimately dwindle down to a few. Then the strategy becomes crystal clear.

Let's say there's just 1 marble left in the jar, and it's my turn. I win by taking the last marble.

I win with 2 marbles, also, because I can take both.

But 3 marbles are bad. I have to leave either 1 or 2 marbles for the other player, and these totals give him an easy win.

Four and 5 marbles are good. They empower me to stick my opponent with the jinxed number of 3.

The pattern is a piece of cake. Every number of marbles that divides by 3 is a losing number. That means 3, 6, 9, 12, . . . are bad when it's your turn. All the other numbers (1, 2, 4, 5, 7, 8, . . .) are good.

What does this mean in actual play? We start with a large but known number of marbles. Divide by 3. If it divides evenly, it's an unlucky number. You don't want to go first. Should the other player offer a coin flip to decide who starts, you can "generously" let him start.

When the number of marbles is lucky, and you move first, you have this winning strategy: With each move, simply make sure to leave the other player an unlucky number of marbles. So if you're starting with 304 marbles (which is lucky), you remove 1, leaving an unlucky 303. Keep doing that on every turn, and eventually he'll be left with 3 marbles. That will set you up for a win.

This strategy is foolproof in that it doesn't matter how the other player plays (as long as he doesn't throw a tantrum and knock over the jar). He has to remove either 1 or 2 marbles from his unlucky number. That always leaves you with a lucky number for your next turn.

The other possibility is that the number of marbles is unlucky on your first move. You're doomed if the other player adopts the strategy above. Look on the bright side. Your opponent may not know the strategy, or he may screw up. Someone playing without a strategy is almost certain to leave you a lucky number sooner or later since two-thirds of the whole numbers are lucky. Someone who knows the optimal strategy but slips up, even once, will permit you to win (as long as you don't slip up yourself).

The question asks whether there's a way to predict who will win. There is, when both players are perfect game theorists. Figure out whether the initial number of marbles is lucky. A lucky number means the first player to move will win; otherwise, the second player will.

The outcome is harder to predict with real humans. Even when both players know the correct strategy, the chance of errors increases with the initial number of marbles. The odds favor the player who's more accurate in applying the strategy.

Interviewers also ask a variant of the question in which the player who takes the last marble loses. In that case, it's numbers of the form $3N + 1$ that are unlucky, and the same basic strategy applies.

? You've got a fleet of fifty lorries, each with a full tank of petrol and a range of 100 miles. How far can you deliver a payload? What if you've got N lorries?

Some have trouble grasping the concept. It's a postapocalyptic world with no petrol stations. The only petrol is whatever is in the lorries' tanks. You can't trade in the lorries for Priuses. It's okay to abandon lorries out in the middle of nowhere. Drivers are dispensable, too. The only thing that matters is getting that precious payload down the road.

There's enough fuel to take fifty lorries 100 miles each. That should be enough petrol to take one lorry 50×100 or 5,000 miles, but is 5,000 miles the answer? Not unless you've got a way to teleport petrol from one lorry's tank to another. Remember, each lorry already has a full tank, so more petrol can't be added until the existing petrol is used up.

Start simple. Imagine there's just one lorry instead of fifty. Put the payload in the flatbed, hop in, and drive. You'll conk out at 100 miles.

Now suppose there are two lorries. Put the payload in the first and drive 100 miles. Can the second lorry help? Not now. It's 100 miles behind you. It would have to follow the first lorry's route and reach it just as its own tank was running dry.

Maybe the first lorry should tow the second. When the first lorry runs out of petrol, unhitch it and climb in the second lorry, with its still-full tank. That's good for another 100 miles.

How far would the first lorry have gotten? Not 100 miles. It was carrying twice the normal weight. Physics demands that it could have gone only half the distance at most—and that's the best-case scenario. Realistically, vehicles' fuel efficiency decreases even more when towing a lot of weight.

There's another approach. Have the two lorries set out simultaneously and travel apace. At 50 miles, each lorry's tank will be half-full, making a full tank of petrol between them. Siphon the petrol from one tank to the other. This gives one lorry a full tank. Abandon the empty lorry and drive the full one another 100 miles. The total mileage is 150. Unlike the mileage in the towing case, this isn't a theoretical limit. It's completely practical.

With three lorries, towing would be a dubious proposition, but the siphoning idea still works fine. Have three lorries start out in tandem. They stop at a third of the 100-mile range, at 33 1/3 miles. Each then has 2/3 of a tank of petrol. Sacrifice one lorry by siphoning its petrol into the two remaining lorries, giving them both full tanks. This exactly reproduces the start of the two-lorry case. As we already know, that's good for 150 miles. With the 33 1/3 miles, that comes to just over 183 miles total.

By this point, the pattern is evident. A lone lorry can go 100 miles. Adding a second lorry gives an extra $100/2 = 50$ miles. A third lorry ups the total by 100/3 miles, a fourth lorry adds 100/4 miles to that. For N lorries, the total mileage is

$$100 * (1/1 + 1/2 + 1/3 + 1/4 + 1/5 + \cdots + 1/N)$$

The fractions part of this is known as the harmonic series. I'll say more about that in the answer to another question. The harmonic series is easily calculated. When $N = 50$, the sum of the series is 4.499+. Multiply that by 100 miles, and you'll see you can drive the payload 449.92 miles.

As N increases, so does the series' sum. With enough lorries, you could drive the payload as far as you liked. However, the distance grows ever more slowly with increasing N, and the energy efficiency becomes ridiculously bad. The thousandth lorry would add only 1/10 mile to the payload distance (but spew out just as much CO_2 as the other lorries, hastening the assumed apocalypse). The millionth lorry would add only inches.

The above is the answer that interviewers are generally looking for. There's an arguably better answer *if* there's a way of carrying fuel and *if* the payload isn't too heavy (you could ask your interviewer about these points).

The question talks about lorries, and lorries are not cars. By definition, they are engineered to carry a big, heavy cargo. Your basic flatbed lorry has a curb weight of about 5,000 pounds and a load capacity of another 5,000 pounds, give or take. It's designed to carry those 5,000 pounds safely, assuming you're not hauling 5,000 pounds of packing peanuts or cotton candy.

A lorry's fuel tank holds about 30 gallons. A gallon is four quarts, a quart is roughly a liter, a liter is a kilogram of water, and a kilogram is about two pounds. (Fermi question wonks would know everything in the previous sentence.) A gallon of water is therefore about 8 pounds. Petrol weighs about 0.75 as much as water, so call it 6 pounds to the gallon. Multiply by 30, and that's not quite 200 pounds for a lorry's tank of fuel.

The key point is that a lorry's fuel weighs way, way less than the lorry itself. It's about 200/5,000 or 1/25 of the lorry's curb weight.

It's crazy to tow or drive a 5,000-pound lorry when all you care about is the 200 pounds of petrol in its tank. It would be better to carry petrol in the flatbed of the lorry with the payload. (Maybe you can salvage containers for the petrol, or tear apart the other lorries and use their fuel tanks as containers.) A lorry could handle about twenty-five lorries' worth of fuel, assuming that the payload doesn't weigh much.

That means that a single lorry could carry half the fuel of the fifty-lorry fleet. It could go something like 25 × 100 or 2,500 miles. Okay, it wouldn't go that far because the load would decrease the fuel efficiency. It still ought to be good for about

1,500 miles. That's more than three times the 450 miles of the siphoning answer, and it requires only one lorry and one driver.

? Simulate a seven-sided die with a five-sided die.

An octahedral die is used in the Magic 8 Ball. Dungeons & Dragons uses dice shaped like all the Platonic solids (with four, six, eight, twelve, or twenty regular-polygon faces). It is more of a challenge to design a fair five-sided die. U.S. Patent 6,926,275 went to one design with two triangular faces and three rectangular ones. The edges are beveled to prevent chipping. When the die lands on a rectangular face, you read the value from the upper number on either of the visible rectangular faces. The roll in the picture would count as 3.

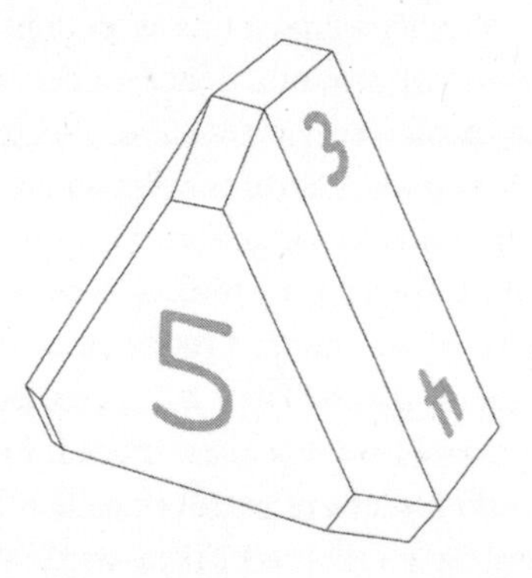

So in case you're wondering, there *is* such a thing as a five-sided die. The question essentially says that you're given a device that produces a random number in the range of 1 to 5. You have to use it to produce a random number in the range of 1 to 7. Pretend there are seven argumentative people holding seven lottery tickets, #1 through #7. How do you use a five-sided die to pick a winner, knowing that the losers are going to complain and that you may have to prove in court that the procedure was truly random?

Several simple ideas are unfair to somebody. One is to roll the five-sided die twice and add the numbers. This produces a number in the range of 2 through 10. It may look like we're making progress, but not really. Any craps player knows that not all dice totals are equally common. Totals in the middle of the distribution (like 7) are more probable. The same applies to five-sided dice.

Another idea is to roll the die twice and total the numbers, or multiply them, or otherwise generate a large number. Then divide by 7, taking only the remainder. The remainder will be in the range of 0 to 6. We don't need a 0, so pretend it's a 7. That gives a "random" number in the range of 1 to 7.

I put "random" in scare quotes because, as the mathematician John von Neumann wrote, "Any one who considers arithmetical methods of producing random digits is, of course, in a state of sin." While this trick may be good enough for some purposes, the result isn't truly random, and so this answer is not rated highly at Google or Amazon. On the web, random numbers had better be random. Otherwise, hackers can take advantage.

For an authentically random outcome, let each of the seven lotto players roll a five-sided die once. The player whose number is highest wins. If there's a tie, roll again (as many times as needed). The only catch is that there's a lot of die rolling. Even with no ties (and ties will be common), there are seven rolls.

There is a better answer. Think digital. The numbers 1 through 7 can be represented in three bits, as binary numbers from 001 to 111. Can you generate three random bits using a five-sided die?

Sure. Each roll will give one digit of the three-bit number. If the die comes up 2 or 4, call the result 0; if it comes up 1 or 3, call it 1; and if it comes up 5, roll again. Keep rolling, as needed, to get a non-5.

Doing this procedure three times generates a number in the

range of 000 to 111. Translate back into decimal, and the person holding that number wins (i.e., 101 means lottery ticket #5 wins). Should 000 come up, you try again.

This takes just three rolls (if no repeats are necessary). On average, it's a little more than four rolls.

? You have an empty room and a group of people waiting outside the room. A "move" consists of either admitting one person into the room or letting one out. Can you arrange a series of moves so that every possible combination of people is in the room exactly once?

It may take a moment to understand what the interviewer wants. Say there are two people outside the room, Larry and Sergey. Then there are four possible combinations of people inside the room—counting the case where the room is unoccupied. The four cases are as follows:

No one in the room
Just Larry in the room
Just Sergey
Larry and Sergey

The question is whether we can start with the empty room and run through each of these combinations. Only one person can enter or leave at a time, and no combination can be repeated, even for a split second. The order of the above list wouldn't work because there's no way of getting from "Just Larry" to "Just Sergey" in one step. Either Larry leaves before Sergey enters, in which case we repeat the "No one" case, or Sergey darts in just before Larry leaves, in which case there's a moment where both are in the room. Here is the solution:

1. Start with no one in the room.
2. Put Larry in the room.
3. Add Sergey, for Larry and Sergey.
4. Larry exits, leaving just Sergey.

This is a simple case, and you're asked to scale up to a possibly large number of people, *N*. Each person may be in the room or out of it, meaning that the number of combinations grows exponentially with *N*, making it unrealistic to play it by ear. You need a good algorithm.

There are two common ways to solve this puzzle. One is to start small and build up. We know how to solve the problem with two people. Suppose we add a third person, Eric. How does that change things? It basically means we have to repeat the two-person directions twice, once without Eric and again with Eric in the room.

Start the same way as above:

1. Start with no one in the room.
2. Put Larry in the room.
3. Add Sergey, for Larry and Sergey.
4. Larry exits, leaving just Sergey.

Then bring in Eric:

5. Eric enters, joining Sergey.

We want to repeat the original directions, with Eric on hand. But we have to repeat them backward, as we're starting where the original ended (number 4), with Sergey alone in the room. Essentially, we're rewinding the Larry-and-Sergey directions. Every entrance becomes an exit and vice versa. Eric remains in the room throughout. Here are the rest of the instructions:

6. Larry joins Sergey and Eric in the room.
7. Sergey exits, leaving Larry and Eric.
8. Larry exits, leaving just Eric.

This sets the pattern for an algorithm. To deal with a fourth person, you run these eight directions, bring in the fourth person, and rewind. It thus takes sixteen steps to deal with four people. The number of steps doubles with each additional person. For n people, 2^n movements are needed.

In the broadest sense, this question is about the clash between the analog and the digital. Moving people in and out of a room is an analog process. You can't blink a human being from one place to another as easily as flipping a digit. Such issues date from the beginning of the information age. Frank Gray was a scientist at Bell Labs in the years when it was a prime mover of the digital juggernaut. Gray developed many of the principles behind color television. His name is best remembered for the Gray code, devised in the mid-1940s.

Early television was strictly analog. A horizontally scanning electron beam was deflected up or down by a magnetic field created by an ever-changing voltage. Gray wanted to convert the analog voltage into a digital number (a series of coded pulses). Engineers of the time had the rather steampunk notion of shooting the electron beam through a mask with holes representing binary numbers. Different parts of the mask, corresponding to different degrees of deflection, had different patterns of holes. The beam would spell out the correct voltage in binary numbers. Like many clever ideas, it didn't work. Electron beams are messy. It was like shooting a squirt gun at a misbehaving cat.

The real problem was that this produced erratic readings

when the beam moved from one voltage number to another. To make it work, Gray needed a number code where only one digit changes from each number to the next. Such systems are now called Gray codes. You can have them with any base, including base 10, but the best-known example is the binary Gray code. It looks like this:

Number	*Gray Code*
0	000
1	001
2	011
3	010
4	110
5	111
6	101
7	100

The digits in Gray's code do not represent powers of 2, or anything really. It's simply a code. The code 111 happens to mean 5, and you shouldn't try to read anything more into it. The sole reason for the existence of the Gray code is that each number can be generated from its predecessor by changing exactly one digit. To get from 5 (111) to 6, you just change the middle digit (to 101).

Gray supplied a simple procedure for generating his codes. Start with 0 and 1. These will be assigned to the regular numbers 0 and 1 (no hocus-pocus there). Then reverse that 0, 1 sequence—to 1, 0—and append it to the original. This gives 0, 1, 1, 0.

In order to distinguish the original sequence from its reversal, we have to add an extra digit to the left of each code. Use 0 for the original sequence and 1 for the reflected version. This gives us 00, 01, 11, 10.

These are the first four Gray codes. Want more? Reverse this sequence and tack it onto the original: 00, 01, 11, 10, 10, 11, 01, 00. Then add a 0 to the first four codes, and a 1 to the last four: 000, 001, 011, 010, 110, 111, 101, 100.

That is why 6 can be represented as 101. You can see, without too much trouble, that the number 8 would have the code 1100. Gray's scheme is readily extended as far as you care to go.

Gray codes are cyclical. Imagine you managed to drive your car a million miles. The odometer reads 999999, then changes to 000000 (there's no millions digit.) With Gray codes, the last number also returns to the first number, but just by changing a single digit. In the table above, the highest number (100) can be changed to the lowest number (000) just by flipping one bit.

A Gray code can be used to solve this interview puzzle. Any engineer given this question is expected to make that connection.

Represent the state of the room as an n-digit number, where n is the number of people. Each digit corresponds to a different person. The digit is 1 if that person is in the room and 0 if he or she is outside. Here's an example:

Stu	*Ann*	*Emily*	*Bob*	*Phil*	*The Other Phil*	*Lisa*	*Eric*	*Sergey*	*Larry*
1	1	0	0	1	0	0	0	0	1

Every possible n-digit binary number (2^n of them) represents a different grouping of people. We need to cycle through *all* of them. The usual counting order of binary numbers won't do. But Gray codes work fine. Just run through the Gray codes in order, starting with 0000000000, and interpret them as stage directions. (For instance, a switch from 0 to 1 in the rightmost place means "Enter Larry.") The solution starts like this:

0000000000: The room is empty.
0000000001: Enter Larry.
0000000011: Sergey joins Larry.
0000000010: Larry leaves.
0000000110: Eric joins Sergey.

It's guaranteed that only one digit changes with each step, and exactly one person enters or leaves the room.

The Gray code is a skeleton key to many classic puzzles, notably the Tower of Hanoi and Chinese Rings. The names may not be familiar, and the Asian provenance is bogus, but you've probably seen the puzzles in real or virtual form. The Tower of Hanoi consists of eight cylindrical disks threaded on one of three poles. The player has to move all eight disks to another pole, with the constraint that no disk can be placed on top of a disk smaller than itself. The Tower of Hanoi has become a cliché of puzzle-genre video games (such as *Mass Effect, Zork Zero,* and *Star Wars: Knights of the Old Republic*). All computer science students learn about Gray codes, and a popular assignment is to write code for a Tower of Hanoi game (which they go on to recycle in a video game).

? You've got an unlimited supply of bricks. You want to stack them, each brick overhanging the one beneath it. What is the maximum overhang you can create?

Imagine placing a brick on the edge of a table, overhanging it by an inch. The brick is stable. Push it over another inch, then another, then another. Your intuition tells you that the brick will tip over when it extends by more than half its length.

In more general terms, the brick's center of gravity must rest on something solid. Or in the limiting case, the center of gravity must at least perch precariously on the edge of something solid. For a uniform brick, the center of gravity is in the exact middle, halfway from either end. That's why you can get a maximum overhang of half a brick length with one brick. It would be safer to settle for a bit less than that—but half a brick is the limit.

You have unlimited bricks and are allowed to stack them. Let's try two bricks. Most people will want or need to draw a diagram on the whiteboard. The dots are centers of gravity (CG). The upper dot is the CG of the top brick. The top brick is okay, so long as the bottom brick is stable, because the top brick's CG is (barely) sitting on the bottom brick.

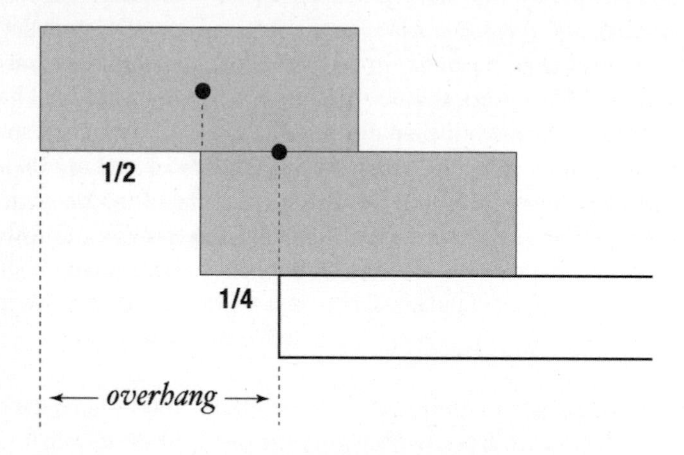

How far can the bottom brick overhang the table? The center of gravity of the two-brick stack (lower dot) must rest above the table or its edge. If it's over air, the stack will topple. This combined center of gravity is a quarter of a brick length to the right of the CG of the top brick alone. We thereby have $1/2 + 1/4 = 3/4$ brick lengths' overhang with two bricks.

Interviewees feel pressure to arrive at an answer quickly. It may now look like each brick's overhang will be half that of the one above it. Since $1/2 + 1/4 + 1/8 + 1/16$ begins an infinite series that adds up to 1, some candidates say you can get an overhang of 1 brick length. Indeed, you can build a stable stack where the overhangs halve with each brick from the top. This isn't the optimal design, though, and this answer won't win you many points from an interviewer.

Here's an analogy that will point you in the right direction. A smart kid transfers into a class. How much is she going to raise the class's average grade? The bigger the class, the less difference the smart kid will make. Her effect is proportional to $1/n$, where n is the number of students. It also depends on how much smarter the new kid is. A new kid who happens to have the same grades as the class as a whole wouldn't change the average at all. The smart kid's effect is proportional to the difference between her grade and the class's average grade.

The same reasoning applies to the bricks. A center of gravity is a (literal) weighted average. Look at the nth brick. We're going to lift the whole existing stack—gingerly—and position it so that its CG is right over the new brick's edge. How far can the new brick overhang the table?

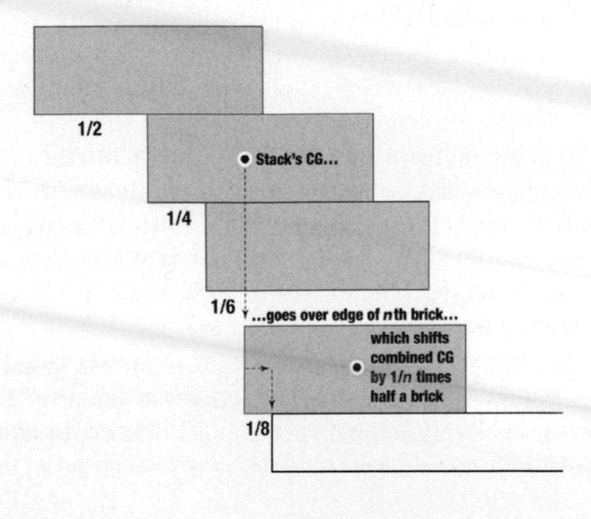

To find out, you have to recompute where the stack's CG is (incorporating the new brick). The new brick's CG is of course in the middle of that brick, half a length to the right of its left edge, and the left edge is where you positioned the CG of all the bricks above it. The new brick's mass "pulls" the collective CG to the right. Its effect depends on the number of bricks and the distance between the current stack's CG and the new bricks' CG. With the above placement rule, the horizontal distance is always half a brick. With n bricks, the new brick has $1/n$th the weight of the whole stack. That means that the nth brick shifts the stack's CG $1/n \times 1/2$ a brick length.

The overhangs therefore form this series:

$$1/2 + 1/4 + 1/6 + 1/8 + 1/10 + 1/12 + \cdots$$

Were you to double that, you'd have this easy-to-remember series:

$$1/1 + 1/2 + 1/3 + 1/4 + 1/5 + 1/6 + 1/7 + \cdots$$

This happens to be a famous series in mathematics and music theory. It's called the harmonic series because the overtones of a vibrating string have wavelengths of 1/2, 1/3, 1/4, 1/5, and so forth, relative to the fundamental tone.

In music, the harmonic series is associated with rich, buttery tones. In maths, it's somewhat notorious. Since each term gets smaller, you might think the harmonic series would converge to a tidy sum (the way that $1/2 + 1/4 + 1/8 + 1/16 + \cdots$ converges to 1). Instead, it sums to infinity.

Applied to bricks, this produces a paradox. A stack's maximum overhang equals half the sum of the harmonic series, and half of infinity is infinity. That means that you can achieve any overhang you like (theoretically). A tall stack of bricks could span the Golden Gate!

A tough interviewer may ask you to prove that the series has no limit. There's an easy way to do that. Nicole d'Oresme, one of the geniuses of the Middle Ages, came up with it. Group the terms of the harmonic series into parentheses enclosing one, two, four, eight,... terms.

$$\left(\frac{1}{2}\right)+\left(\frac{1}{3}+\frac{1}{4}\right)+\left(\frac{1}{5}+\frac{1}{6}+\frac{1}{7}+\frac{1}{8}\right)+\left(\frac{1}{9}+\frac{1}{10}+\frac{1}{11}+\frac{1}{12}+\frac{1}{13}+\frac{1}{14}+\frac{1}{15}+\frac{1}{16}\right)+\cdots$$

Then write this series beneath it.

$$\left(\frac{1}{2}\right)+\left(\frac{1}{4}+\frac{1}{4}\right)+\left(\frac{1}{8}+\frac{1}{8}+\frac{1}{8}+\frac{1}{8}\right)+\left(\frac{1}{16}+\frac{1}{16}+\frac{1}{16}+\frac{1}{16}+\frac{1}{16}+\frac{1}{16}+\frac{1}{16}+\frac{1}{16}\right)+\cdots$$

Now compare. The sum of the harmonic series (top) is certainly bigger than the sum of the series below it, as all its terms are at least as big, and many are larger.

What is the sum of the series on the bottom? Each set of parentheses encloses fractions that add up to one-half. The bottom series is equivalent to an infinite string of one-halves. That's infinite, and since the harmonic series is bigger in side-by-side comparison, it must be infinite, too.

All well and fine. But if you end there—claiming an infinite overhang—you might be accused of giving a technically correct though completely useless answer. You can impress the interviewer by showing an ability to bridge the theoretical and the practical. Suppose the *Guinness Book of Records* folks heard about your "infinite overhang" claim and wanted to shoot a video. How much overhang could you actually deliver?

That is something quite different, and fortunately, it's easily estimated from the fact that the harmonic series grows incredibly slowly. As we've seen, four bricks yield almost a brick length of overhang. Five bricks just exceeds it, leaving the top brick levitating completely beyond the table edge. That's an easy parlor trick.

(If you don't have any bricks handy, you can make do with dominoes, books, CD cases, etc.)

With ten bricks, the maximum overhang is just over 1.46 brick lengths. With a hundred bricks, it's about 2.59, and for a thousand bricks, it's 3.45. Given vibrations, wind, and the imperfections of bricks, it seems unlikely that anyone could stack a thousand bricks in a straight unmortared column, much less an overhanging one. Conclusion: an overhang of two brick lengths is doable; three is dicey and dangerous.

This riddle has more connection to the realities of bricklaying than you'd probably think. In architecture, a corbel arch consists of two overhanging brick stacks meeting to form an arch in the middle. The Mayans were building them as early as 900 BC. There's one problem with corbel arches: the buildings tend to fall down. The so-called Roman arch with a keystone, known in ancient Mesopotamia, gradually supplanted corbel arches worldwide.

The brick-stacking puzzle appeared in an 1850 engineering text, *Elementary Mechanics*, by J. B. Phear. It was discussed by the physicist George Gamow in a 1958 book, *Puzzle-Math*, and later by Martin Gardner. Variations of it (allowing more complicated stacks, with more than one brick per horizontal row) have inspired serious mathematical articles. The interview question about a fleet of fifty lorries is closely related. It's analogous to asking how much of an overhang you could get with fifty bricks.

? You have to get from point A to point B. You don't know whether you can get there. What do you do?

The MBA answer: "I would pull out my mobile phone and enter point A and point B in Google Maps. If point B isn't on Google Maps, I'd take a taxi and submit the receipt to accounting. Next question?"

The computer science PhD answer: "Oh, I get it. You're asking about the problem of searching a network. . . ."

When asked of a software engineer, this question is expected to lead to a discussion of the relative merits of specific search algorithms. Though these algorithms were devised for searching computer memories and the Internet, they are equally relevant to navigating a shopping center, a garden maze, or the quaint villages of Umbria. I will give a commonsense answer that is ultimately not so far from the computer scientist's.

To rephrase the question, you're at point A, you want to find point B, and there's no app for that. You have to stick to the roads or paths leading from A. You'll recognize point B when and if you come to it. But you might *not* come to it. Point B could be off the road network and unreachable.

You should begin by asking several important questions of the interviewer:

1. Can I ask for directions? Can I use GPS? Do I have any way of estimating the direction of or distance to point B?
2. In the event that point B is *not* reachable from A, is there any way of realizing that, short of a never-ending search?
3. Am I interested in finding point B as quickly as possible, or in finding the fastest route from point A to point B as quickly as possible?

Interviewers like to hear that you're smart enough to ask for directions, but you'll be told that you can't count on getting foolproof guidance. Question 2 is important because, well, the better engineers like to avoid throwing infinite time and effort into a bottomless pit. You don't want to search the whole planet to determine that you can't get there (B) from here (A).

That last question, 3, may be a little confusing. To give one example, you might be lost in a cornfield maze at point A with two screaming kids. You want to find the exit, point B. All you care about is getting out of the frigging maze.

You'd want a search procedure that would find point B with due dispatch. However, there are almost always false turns, and the route you'd take wouldn't necessarily be the very shortest route from A to B. In this situation, that would be okay.

Alternatively, maybe you want to commute from home (A) to work (B) on public transportation. You will be repeating the route you find every working day of your life. You are not just looking for B; you're looking for the shortest route between A and B.

Any search has an element of trial and error. There is usually an element of knowledge or intuition, too. You may have beliefs on how to get to B, based on maps, hunches, street smarts, the wisdom of French Canadian trappers, or a road sign saying POINT B 17 MILES. A search procedure should marshal whatever information you have (and allow for the possibility that this information is not reliable). You'll begin by exploring the route that you believe is most likely to be the shortest route to B. Make a map as you go, in case you have to backtrack and try different routes.

So far this is incontestable. To impress the interviewer, you'll need to say something not quite so obvious. Try this: the fundamental philosophical question of searching for a destination is, "When should I turn back?"

There may come a time when you feel you are lost—meaning that you believe you have strayed far from the most direct path from A to B. Do you backtrack to where you were before you were lost? Or do you try to find the most direct path from where you are now (lost) to point B?

There's a good chance you've heard this issue debated on your last long trip. If the jokes about male drivers are correct, men are loath to backtrack or ask for directions. Suppose a friendly stranger assures Ashley and Ben that point B is just down that road, saying,

"You can't miss it." They drive half an hour, expecting every moment to glimpse B around the next bend. It never happens. "We're obviously not on the right road," Ashley announces. "Let's go back to where we were before we got those directions."

"There's no point in turning back," counters Ben. "We've come a long way, and we must be closer to B than we were. There's got to be a sign up ahead."

Ben's strategy resembles what computer scientists call the best-first algorithm. Whenever you come to a fork in the road, you follow the presumed shortest route to B, based on your current state of knowledge. In the happy case where this knowledge is 100 percent accurate, Ben will make a beeline for point B.

Ashley's strategy is more like the A* search algorithm (pronounced "A star"), described by the computer scientists Peter Hart, Nils Nilsson, and Bertram Raphael in 1968. This says (roughly) that you should stick as close as possible to the shortest path from A to B. You're probably wondering how that's any different from Ben's strategy. It's not, as long as the searcher has reliable directions. The difference arises when the searcher strays from the direct route. In deciding what to do, Ben looks at a single number: his guesstimate of how far B is from his present location. He always tries to move in the direction of B. Ashley looks at two numbers: her estimated distance to B and her known road distance from A. Ashley's goal is to minimize both numbers or, more exactly, their sum. Ashley tries to explore the points that are most likely to lie on the shortest route from A to B.

Who's right, Ben or Ashley? Ashley's search procedure is better at dealing with wrong turns. The diagram shows the gist of the matter. Starting out from A, the searcher comes to a fork in the road and must choose either the right or the left path. Should Ben choose the left path—a mistake!—he will endure a lengthy trudge. Though long and circuitous, this route takes him ever closer to B.

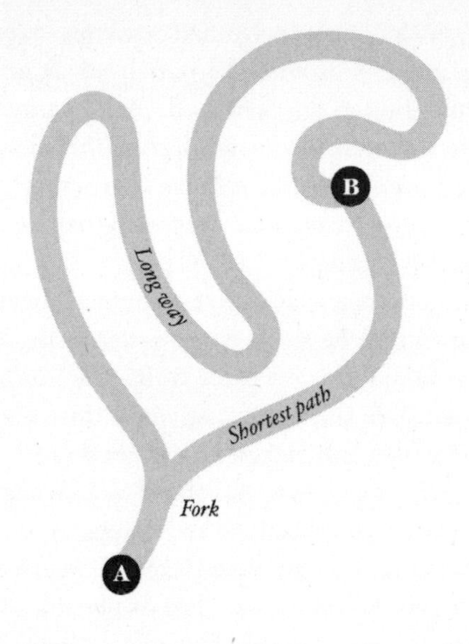

Should Ashley take the same wrong turn, she would eventually realize that she's come an awfully long way from A, and B is still far away. This would tell her she's not likely to be on the shortest path. Ashley would backtrack to the fork and try the other path. She'd probably find B quicker than Ben would. In general, you'd expect there to be many forks, rather than just one, and decisions all along the way. Similar trade-offs apply: a search method with an optimal propensity to backtrack beats one that backtracks too infrequently.

This interview question further slants things in favor of an A*-like search by saying that you don't know whether you can get from A to B. If there's no way of getting to B, Ben will wander aimlessly, forever chasing a will-o'-the-wisp. Ashley will explore systematically outward from point A, as she's minimizing distance from A. She will form a map of the territory that will help to establish that there is no way of getting from A to B. That will prevent wasting further resources.

A* search has particular merit when the goal is to find the shortest path of all. For that reason, it's used in mapping apps and in video games, to route characters through their virtual worlds. A*-like searches arguably have psychological advantages, too. As fallible humans, we are prone to self-justifying convictions. It's seductively easy to burn resources exploring the wrong route, the wrong business plan, the wrong romantic partner, the wrong idea—all the while convinced that success is just around the next corner. An A* search is able to pull the plug on an unprofitable venture and begin anew somewhere else. That's an idea of wide relevance. Success is about not giving up too easily but also about knowing when to throw in the towel.

Bottom line: the best way to get from point A to point B is to stick close to the route you currently believe is shortest (an A* search)—that, rather than focusing just on finding point B.

? How do you find the closest pair of stars in the sky?

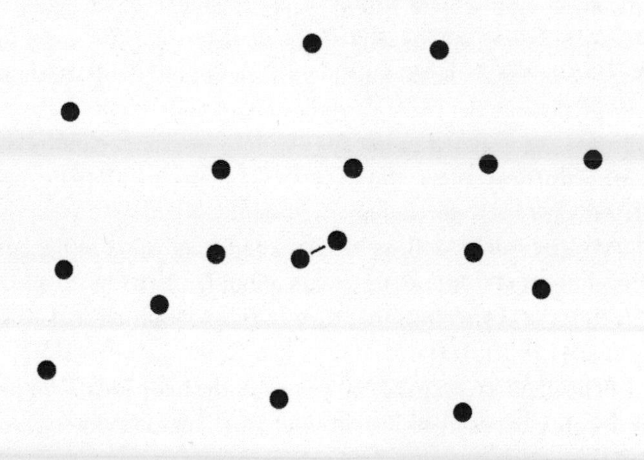

This is the "closest pair" problem, well known to computer scientists. The human eye can often spot the closest pair of stars (or points in a plane) at a glance, just as the eye can tell whether a face is male or female or solve a Captcha. Microprocessors can't readily do any of these things. Algorithms for solving the closest-pair problem were intensively studied in the 1970s and are now a building block of the digital world. Even your smartphone has apps that use them, for purposes ranging from mapping to games to cameras. It's part of how our clever devices are starting to "see."

This means that the interviewer is really asking a technically trained candidate, "Were you paying attention in algorithm class?" More specifically: given the coordinates of a large number of random points in a plane, how can a computer, with no visual cortex, tell which pair is closest by pure computation?

One inefficient approach is to compute the distance between every pair of stars. That's a lot of calculations when N is a large number.

But it's not actually necessary to check the distance between *every* pair of stars, only those that are "reasonably close." Unfortunately, computers are stupid. They can't tell which pairs are "reasonably close" unless they do the maths.

The optimal closest-pair algorithm goes like this. Mentally split the sky in two. There's a right half and a left half, each with $N/2$ stars. Keep partitioning the sky into quarters, eighths, sixteenths, thirty-seconds, and so forth. The cuts are all to be made vertically (or from celestial north to south).

A suburbanite will see maybe a thousand stars in his hazy, light-polluted sky. It therefore takes about ten halvings to end up with strips of sky so thin that they contain about two naked-eye stars each (2^{10} is 1,024).

The diagram on page 250 gives you the basic idea. Compute the distance between each strip's star pair. That's vastly less work than computing distances between all the stars.

A complete doofus might think that you're almost done. Find the strip with the closest pair of all, and that's it! Hardly—

look at the diagram. Because the strips are so long and narrow, the closest pair within a strip may not be so close at all. The closest pair in the whole sky is likely to be two stars straddling two strips. If you look below, I've circled a straddling pair in the diagram.

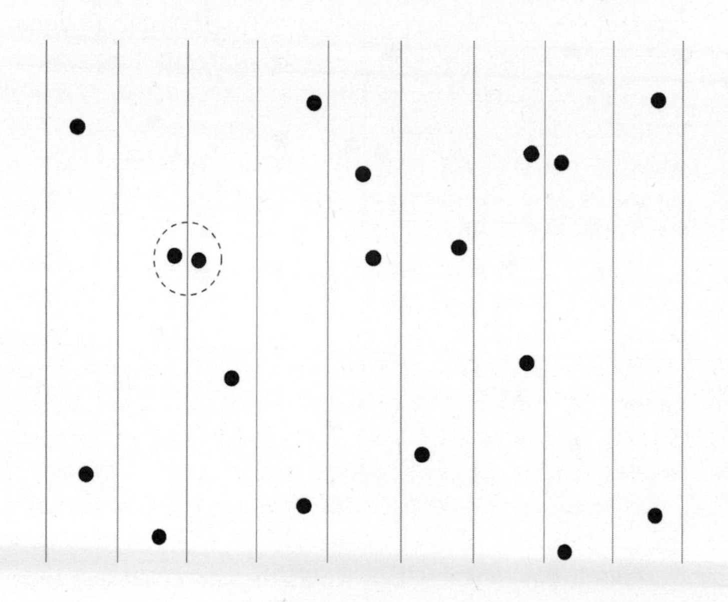

It's necessary to have an algorithm for knitting together two adjacent slices. You give the algorithm the closest pair in the left slice and the closest pair in the right slice, and then it deduces the closest pair in the combined left-plus-right slice. This algorithm would then be applied, over and over, to create slices twice as wide, four times as wide, eight times, and so forth. After ten stages of knitting together, we'd be back to a "slice" consisting of the whole sky. We'd also know the closest pair in the whole sky.

The knitting-together algorithm works like this. Given two adjacent slices, inspect a zone centered on the dividing line to see whether it's got a pair closer than any in either half. If it does, then that will be the closest pair in the combined slice.

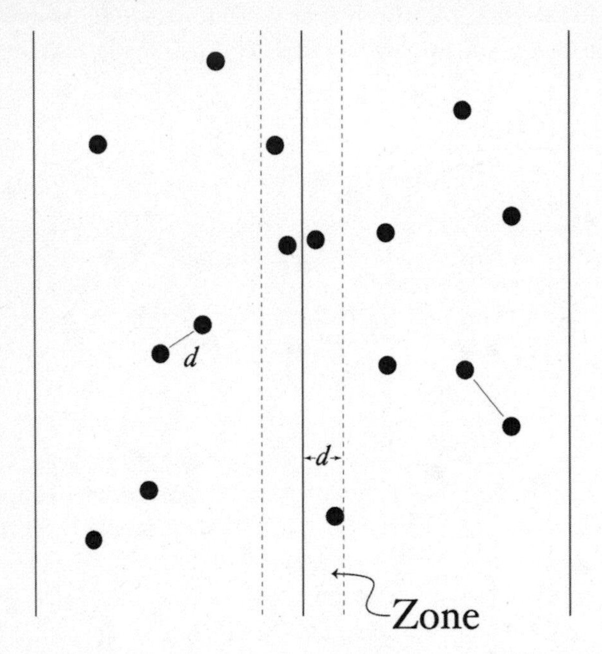

Take the closest pair wholly within the right or left slice and call the distance between its stars *d*. Think of *d* as the distance to beat. We need to search within distance *d* of the dividing line for possible straddling pairs. The zone is a 2*d*-wide strip.

This can be efficiently done. You might be thinking, "Sure, whatever." Computer scientists have a way of worrying about worst-case scenarios (which the user experiences as bugs). The stars in the zone can be readily sorted by their vertical coordinates. For each such star, it's necessary to check distances to stars within *d* of it, up or down. A simple diagram (facing page) proves that there can be only six stars to check at most. That's nicely manageable.

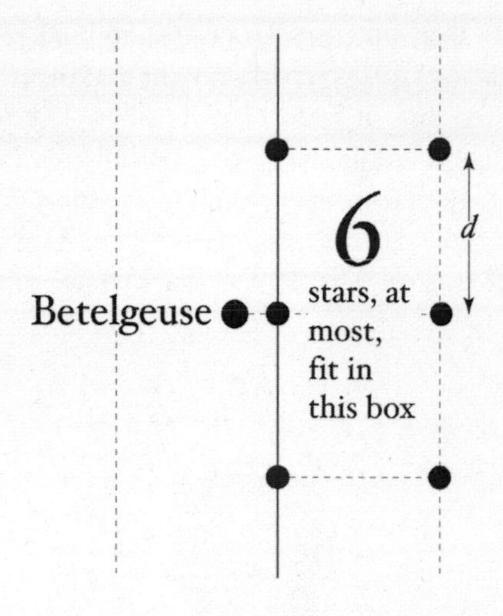

Here we want to make sure that Betelgeuse, just to the left of the dividing line, doesn't form a close line-straddling pair with a star on the right. This defines a $d \times 2d$ box to check. Since we already know that no two stars on the right are within d of each other, there can be only six stars, at the very most, that fit within the box.

If you haven't studied computer science, your head is probably spinning. For coders, this question should be easier than the brainteasers—it simply asks them to recite what they learned at university.

There is an offbeat answer that rates mention. Astronomy buffs may correctly point out that there are two kinds of close star pairs. They are *binary stars* (two stars orbiting around each other, the way the earth orbits the sun) and *optical doubles* (unrelated stars that just happen to look close in the sky, as seen from the earth). Given the profound emptiness of space, it's all but a certainty that the closest pair of stars in the earth's sky will be a binary.

Not only that, but there are eclipsing binaries, where the orbital plane is so close to our line of sight that the two stars actually pass behind and in front of each other. The most famous example is Algol, a visible star in the constellation Perseus. Algol looks like a single star to the eye and to the most powerful telescopes. Every three days, its brightness varies. Astronomers have determined that Algol is really two stars whose orbits look like this, to scale:

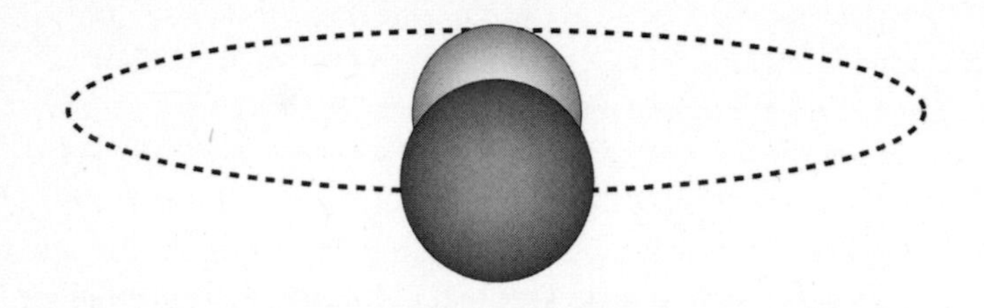

When one star partly blocks the other, Algol looks dimmer, even to the unaided eye. During an eclipse, there is no distance between the two stars' disks. They actually touch and overlap in the sky. That's as close as a pair of stars can get.

How do you find the closest pair? The astronomer's answer is that eclipsing binaries are identified by their regularly varying brightness and spectroscopic signatures.

Chapter Ten

? Can you swim faster through water or syrup?

Isaac Newton and Christiaan Huygens debated this question in the 1600s, without resolving it. Three centuries later, two University of Minnesota chemists, Brian Gettelfinger and Edward Cussler, did a syrup-versus-water experiment. Maybe it's not so surprising that it took so long. Cussler said he had to obtain twenty-two approvals, including permission to pour massive quantities of syrup down a drain. He had to say no thanks to enough free corn-syrup to fill twenty lorries because it was judged hazardous to the Minneapolis sewer system. Instead he used guar gum, an edible thickener used in ice cream, shampoo, and salad dressing. About six hundred pounds of the stuff transformed a swimming pool into something that "looked like snot."

Gettelfinger, an Olympic-hopeful swimmer, had the unique experience of diving in. As to the outcome, I will leave you in suspense for a bit longer. The findings were published in a 2004 article in the *American Institute of Chemical Engineers Journal.* The next year, Gettelfinger and Cussler won the 2005 Ig Nobel Prize in Chemistry. The Ig Nobels are the silly counterpart to the better-known prizes awarded in Stockholm, and they automatically get news-of-the-weird coverage. The media attention was apparently responsible for the syrup riddle's revival as a uniquely sadistic interview question.

In the great syrup swim-off, the guar-gum glop's viscosity was about twice that of water. Its density was virtually the same as that of plain water. That was important because, as swimmers have long known, people swim faster in denser saltwater. Like a ship, a swimmer's body rides higher in saltwater, encountering less drag.

Gettelfinger and other Minnesota students did laps through both water and "syrup." They tried the standard strokes: backstroke, breaststroke, butterfly, and freestyle. In no case did the speeds vary between fluids by more than a few percentage points. There was no overall pattern favoring the glop or the water.

This meant that Newton was wrong: he thought syrup's viscosity would slow swimmers. Huygens correctly predicted there would be no difference in speed. Gettelfinger and Cussler's paper supplies the reason. Think of the way smoke rises from a cigarette. For a few inches, the smoke rises in a sleek vertical column. Above that, it breaks up into complicated swirls and eddies. The swirls are turbulence. Turbulence is bad for jet planes, speedboats, and anything that wants to cut quickly through a fluid. Because the human body is not optimized for swimming, we create a ridiculous amount of turbulence, which we then fight in our struggle to pull ourselves through the water. The turbulence produces much more drag than the viscosity does. Relatively speaking, the viscosity hardly matters. As the turbulence is similar in water and syrup, the swimming speed is about the same, too.

The flow of water is much less turbulent for a fish, and even less so for a bacterium—which *would* swim slower in syrup.

Is this a fair interview question? Cussler told me that a background in computer science "is probably of no use" in answering the syrup question. But he added, "Anyone with sophomore physics should be able to answer it." As a physics graduate, I'm willing to bet that's optimistic. At any rate, most of those confronting this question on a job interview won't know much about the physics. Good responses invoke simple, intuitive analogies to explain why the answer needs to be determined by experiment. Here are four points:

1. *You can't swim in the La Brea Tar Pits.* Some fluids are too thick to swim in. Ask the mastodons in the tar pits. Imagine trying to swim in liquid cement or quicksand. Clearly, swimming is slower in *very* thick fluids than in water. Even though there's more to push against, you swim slower, or not at all.
2. *"Syrup" covers a lot of territory.* The question doesn't ask about pitch or quicksand but rather "syrup." There is maple syrup, cough syrup, chocolate syrup, high-fructose corn syrup, and the kind they sip in nightclubs, with consistencies ranging from watery thin to the crud left in the bottom of an old tin of golden syrup. The question is

unanswerable unless you know the exact type of syrup—*or* can prove that swimming is slower in *every* thicker-than-water fluid.

3. *Cite Darwin.* Suppose there's an optimum level of viscosity in which swimming speed is at a maximum. Is there any reason to believe that H_20 just happens to be at that optimum level for swimming?

   You might say yes, were you an insightful fish. Evolution has streamlined fish to "fit" the way water flows over their sleek bodies. Humans are not very fishlike, and the way we swim is not much like the way a fish swims. Neither people nor our immediate ancestors have spent much gene-pool quality time in the pool—or in rivers, lakes, and oceans. Swimming is something we do, like hang gliding, but we're not built for it. A creature built for doing the Australian crawl would look very different from a human. Edward Cussler noted, "The best swimmer should have the body of a snake and the arms of a gorilla."

   It might not be surprising to find that people could swim faster in something with a different viscosity than water. Nor would it be surprising to find that the speed was the same over a wide range of viscosities.

4. *Swimming is chaos.* The dynamics of liquids and gases is a textbook example of chaos. It is so dependent on the granular details as to defy prediction. That's why aircraft designers need wind tunnels to test their designs. The unstreamlined human body, with its relatively awkward movements in water, only complicates things further. This is a question that needs to be tested with an experiment, using the exact type of "syrup" in question.

Cussler's Ig Nobel Prize acceptance speech boiled all this down to six words: "The reasons for this are complicated."

Postscript

Four questions traditionally posed in job interviews at the consulting firm Accenture:

1. How do you put a giraffe in a refrigerator?
   The correct answer: Open the refrigerator, put in the giraffe, and close the door.
2. How do you put an elephant in a refrigerator?
   The correct answer: Open the refrigerator, take out the giraffe, put in the elephant, and close the door. This question tests your ability to recognize the consequences of your actions.
3. The Lion King is holding an animal conference. All the animals attend except one. Which one?
   The correct answer: The elephant. You put him in the refrigerator. This question tests your memory. You've now got one final chance to prove yourself.
4. You have to cross a river in crocodile country and don't have a boat. How do you get across?
   The correct answer: You swim. The crocodiles are all attending the animal conference. This tests how well you learn from your mistakes.

Acknowledgments

In 2003 I published *How Would You Move Mount Fuji?*, a book built around the brainteaser interview questions popularized by Microsoft. Ever since then, I've received a welcome stream of e-mails from people sharing their interview experiences, good, bad, and outrageous. Those correspondents have kept me abreast of "new" questions and styles of interviewing. This book was inspired by their enthusiasm. As before, many of those who were most helpful must go unnamed.

Edward Cussler not only was informative about the history of the "syrup question" but also introduced me to the Oxbridge tradition of puzzling interview questions. At Google, Todd Carlisle and Prasad Setty were generous in explaining their philosophy of hiring. Thanks to Jordan Newman for coordinating interviews and arranging a tour of the Googleplex.

Special thanks go to Rakesh Agrawal, Adam David Barr, Joe Barrera, Tracy Behar, Kiran Bondalapati, John Brockman, Glenn Elert and students, Curtis Fonger, Terry Fonville, Randy Gold, William Hilliard, Larry Hussar, Rohan Mathew, Katinka Matson, Gene McKenna, Asya Muchnick, Alex Paikin, Kathy Poundstone,

Michael Pryor, Michelle Robinovitz, Christina Rodriguez, Arthur Saint-Aubin, Chris Sells, Alyson Shontell, Joel Shurkin, Jerry Slocum, Jerome Smith, Norman Spears, Joel Spolsky, Noah Suojanen, the staff of the UCLA Research Library, Karen Wickre, and Joe Wisnovsky.

Websites and Videos

CareerBuilder: www.careerbuilder.com. A popular CV-posting site, with job listings and advice.

CareerCup: www.careercup.com. Specializes in technology companies—a good source of current interview questions.

Glassdoor: www.glassdoor.com. Covers finance, technology, and other industries. Glassdoor allows users to post salaries, company reviews, and interview questions.

Google Interview: http://google-interview.com. A digest of posts on interviewing (unauthorized by Google!) and overall the most useful Google-specific site. Many interview questions, both technical and brainteaser, plus rants and first-person experiences.

Hacking a Google Interview: http://courses.csail.mit.edu/iap/interview/materials.php. This was a course taught at MIT by Bill Jacobs and Curtis Fonger, focusing on technical questions for software engineers. The course handouts, available online, are excellent.

Monster.com: www.monster.com. The best-known job site, with feature articles on all phases of job searching.

Stanford's Entrepreneurship Corner: http://ecorner.stanford.edu/authorMaterialInfo.html?mid=1090. In this video of a May 1, 2002, talk at Stanford, Larry Page and Eric Schmidt talk about many aspects of Google culture, including hiring.

Notes

Epigraph

"A hundred prisoners are each locked in a room": This question has appeared on innumerable websites. "Elizabeth G.," a listener of radio's *Car Talk*, submitted it to the show's site, claiming she got it from "a friend of a father of a friend," a certain "Alan B." at Electronics for Imaging, Foster City, California. EFI is known for tricky interview questions. The parody was Poll 248 on the Irregular Webcomic! site. There it appeared with a set of answers (poll results):

The 64th square would have more rice than the entire kingdom.: 695 (16.4%)
The surgeon is his MOTHER.: 493 (11.6%)
You should change your choice to the other door.: 485 (11.4%)
The seventh philosopher starves to death.: 472 (11.1%)
He committed suicide with an icicle.: 466 (11.0%)
16 miles per hour.: 389 (9.2%)
Only if the missionary is also the nun's uncle.: 366 (8.6%)
The first cannibal on the 29th night at midnight.: 346 (8.1%)
Ask him what the other farmer would say is the correct road.: 209 (4.9%)
He adds his own horse, then it's left over at the end.: 183 (4.3%)
He's too short to reach any button above the 10th floor.: 145 (3.4%)
In addition, over thirty Douglas Adams fans e-mailed to say the correct answer was 42.

Chapter One

4 Google was receiving a million job applications: Auletta, *Googled,* 15.
4 "You are shrunk to the height of a penny": The dialogue in this section is a composite of several interviewees' accounts.
6 Google pays the income tax: Bernard, "Google to Add Pay."
9 *On-Line Encyclopedia of Integer Sequences:* www.research.att.com/~njas/sequences/index.html.
10 Milton's suggestion was "googol": See Wikipedia entry for Edward Kasner, http://en.wikipedia.org/wiki/Edward_Kasner.
10 "Sean and Larry were in their office": See "Origin of the Name 'Google,'" http://www-graphics.stanford.edu/~dk/google_name_origin.html.
11 "Have Your Google People": Merrell, "Have Your Google People."
12 "are not warm and fuzzy people": Alyson Shontell interview, May 24, 2010.
12 over half of Google's revenue: Auletta, *Googled,* 286, which cites Google vice president Marissa Mayer as the source.
13 "Imagination is more important than knowledge": This is one bumper-sticker aphorism that's authentic. Einstein said it in a 1929 interview in the *Saturday Evening Post.* See Viereck, "What Life Means to Einstein."
13 "One might assume": See February 8, 2005, post by "nuvem" on www.gamedev.net/community/forums/topic.asp?topic_id=299692.
15 Incredible Shrinking Man fights off a spider: LaBarbera, "The Biology of B-Movie Monsters."
16 Borelli, a contemporary of Galileo's, deduced: Borelli, *On the Movement of Animals.*
17 "You can drop a mouse": Quoted in Vogel, "Living in a Physical World," 303.
18 "For the last year my biggest worry": Auletta, *Googled,* 215–16.

Chapter Two

21 "You're in an 8-by-8 stone corridor": Chris Sells, November 23, 2005, post at www.sellsbrothers.com/Posts/Details/12378.
22 "The personnel interview continues to be": Dunnette and Bass, "Behavioral Scientists and Personnel Management."
22 "Most of the corporate recruiters": Martin, "Confessions of an Interviewer."
22 "In an interview you can tell": July 6, 2004, post on Cohen's blog at www.advogato.org/person/Bram/diary.html?start=111.
23 "Interviews are a terrible predictor": Hansel, "Google Answer to Filling Jobs Is an Algorithm."
23 Thomas L. Peters of the Washington Life Insurance Company proposed: Gunter, *Biodata,* 7.

26 Guilford sliced and diced intelligence: The number began at 120 and was revised upward to 150 and finally to 180. See Wikipedia article on J. P. Guilford, http://en.wikipedia.org/wiki/J._P._Guilford.

27 "Genius is one percent inspiration": See the Wikiquote entry for Thomas Edison, which gives several variants and published sources, http://en.wikiquote.org/wiki/Thomas_Edison.

28 "Creativity is production of something new or unusual": Torrance, *Guiding Creative Talent,* 16.

30 "Why do you think it'll amount to anything this time?": Millar, *E. Paul Torrance,* 51.

30 a slough with a few tufts of good footing: In his 1850 novel *White-Jacket,* Melville called philosophy "a slough and a mire, with a few tufts of good footing here and there." See p. 177 of the 1892 reprint (New York: United States Book Company), available on Google Books.

30 define creativity as the ability to combine novelty and usefulness: Cohen, "Charting Creativity."

31 "Oxbridge questions": Edward Cussler interview, June 16, 2010; Moggridge, *How to Get into Oxford and Cambridge.* See also http://dailysalty.blogspot.com/2007/09/brilliant-interview-questions-how-many.html.

31 One of its legendary engineers, John W. Backus: Lohr, "John W. Backus."

32 the term *software* did not exist: Lohr, "John W. Backus," claims that the word dates to 1958.

32 "They took anyone": Lohr, "John W. Backus."

32 "You have to generate many ideas": Lohr, "John W. Backus."

34 turns up in diverse mathematical contexts: In the 2004 SEC filing for its initial public offering, Google requested permission to raise $2,718,281,828. That's a billion times *e* dollars.

35 Wolfram's single line of code: Pegg and Weisstein, "Mathematica's Google Aptitude." Coincidentally, Sergey Brin was once an intern at Wolfram Research.

36 Page was fascinated by the eccentric inventor Nikola Tesla: Auletta, *Googled,* 33.

36 Page or Brin or both interviewed each candidate: Auletta, *Googled,* 98.

36 "Amid the surreal oddity of it all": Auletta, *Googled,* 98.

37 "the amount of people that would cut off": Comment posted on www.glassdoor.com/Interview/Apple-Interview-Questions-E1138_P6.htm.

37 "Have you ever gone to a Church": www.glassdoor.com/Interview/Apple-Interview-Questions-E1138_P6.htm.

37 When Apple opened its store on Manhattan's Upper West Side: Frommer, "It's Harder to Get a Job."

38 Hewlett-Packard was one of the pioneers: Guynn, "Tech firms try to outperk one another."

38 cribbed from other companies like... Facebook: Auletta, *Googled,* 288.

38 "Our competitors have to be competitive": Auletta, *Googled,* 286.

39 Brin supplied a characteristically quantitative defense: Auletta, *Googled,* 57.

39 outperformed the market by 4.1 percentage points: Edmans, "Does the Stock Market Fully Value Intangibles?," 2.

39 Page recalls a grandfather: Auletta, *Googled,* 33.

40 "We do have a workforce": Guynn, "Tech Firms Try to Outperk One Another."

40 41 percent of new 2010 grads: Warner, "The Why-Worry Generation."

40 "He wanted to know if it covered the cost of filing lawsuits": Peter Bailey, quoted in Laakmann, "Cracking the Technical Interview," 15.

Chapter Three

42 job seekers outnumbered openings six to one: Goodman, "U.S. Job Seekers Exceed Openings."

43 "Most people don't interview very frequently": Agrawal interview, June 8, 2010.

43 "How would you improve it?": Agrawal interview, June 8, 2010.

44 "Everything I said they responded 'Are you sure?' ": Comment posted on CareerCup site, www.careercup.com/question?id=1945.

44 estimate the amount of cash in his wallet: Patterson, *The Quants,* 166.

45 "If you were a cartoon character": www.glassdoor.com/Interview/Bank-of-America-Interview-Questions-E8874_P3.htm.

46 "On a scale of 1 to 10, how weird are you?": Bryant, "On a Scale of 1 to 10."

46 Nordstrom (ranked #53 on *Fortune*'s 2010 list...): See http://money.cnn.com/magazines/fortune/bestcompanies/2010/snapshots/53.html.

46 "Just think about if you got stuck": Page, in a talk at Stanford, May 1, 2002. Video at http://ecorner.stanford.edu/authorMaterialInfo.html?mid=1090.

46 "If you see a history of bad decision-making": Glatter, "Another Hurdle for the Jobless."

47 "Most executive recruiters won't look at a candidate": Isadore, "Out-of-Work Job Applicants."

47 "We're definitely putting people through more paces": Tugend, "Getting Hired."

47 Robinovitz predicts this trend will survive: Robinovitz interview, June 17, 2010.

48 a famous one was done by AT&T: Bray, *Formative Years in Business.*

49 "But the trouble was that, in fact, we could not tell": Kahneman, "Nobel Prize Autobiography."

49 view almost *any* criterion as "unfair": Stone and Jones, "Perceived Fairness of Biodata." Among the questions tested for perceived fairness was "Have you ever built a model airplane?"

Chapter Four

51 "Everyone knows Google's doing a good job": http://sites.google.com/site/steveyegge2/google-secret-weapon.
51 first thirty Google employees received stock: Auletta, *Googled,* 109.
52 "Smart people go where smart people are": http://sites.google.com/site/steveyegge2/google-secret-weapon.
52 "They had a lot of data": Carlisle interview, April 7, 2010.
52 "The founders are engineers": Carlisle interview, April 7, 2010.
52 "I started to delve into things": Carlisle interview, April 7, 2010.
53 "We like people to be very collaborative": Carlisle interview, April 7, 2010.
54 "Please indicate your working style preference": Hansel, "Google Answer to Filling Jobs Is an Algorithm."
54 "One of the things I tested for": Carlisle interview, April 7, 2010.
56 "Google was my first job": Juliette, post dated August 1, 2008, at http://techcrunch.com/2009/01/18/why-google-employees-quit/#ixzz0oUskr1wQ.
56 Auletta termed it "preposterous": Auletta, *Googled,* 49.
56 having to supply high school grades: Auletta, *Googled,* 214.
56 Google ignores "Ivy free" CVs: See www.sfgate.com/cgi-bin/blogs/techchron/detail?entry_id=50641.
56 "Last week we hired six people": Hansell, "Google Answer to Filling Jobs Is an Algorithm."
57 "that someone has done that vetting for us": Carlisle interview, April 7, 2010.
57 "We do go out of our way": Page talk at Stanford, May 1, 2002. Video at http://ecorner.stanford.edu/authorMaterialInfo.html?mid=1080.
57 "finding the people we would ignore": Carlisle interview, April 7, 2010.
57 said to be close to 50 percent: See 2006 speech by Omid Kordestani on YouTube, www.youtube.com/watch?v=ZARPcmuTTXs.
57 "art of rejection": Carlisle interview, April 7, 2010.
58 Corporate Executive Board did a similar but broader study: Tugend, "Getting Hired."
58 "I don't think we should hire this candidate": Carlisle interview, April 7, 2010.
60 "We're not trying to take the human element": Setty interview, April 7, 2010.
60 "In most companies, you as a manager go to finance": Setty interview, April 7, 2010.
60 "We don't know whether our system": Setty interview, April 7, 2010.
61 "the belief of Larry, Sergey, and Eric": Setty interview, April 7, 2010.
61 "In an up market, say the late 1990s": Tugend, "Getting Hired."
61 "There's been gradual erosion": Tugend, "Getting Hired."
61 "you can totally bomb an interview": Carlisle interview, April 7, 2010.

62 "You're supposed to ask open-ended questions": BillR, November 19, 2009, post at http://blog.seattleinterviewcoach.com/2009/02/140-google-interview-questions.html.

62 "What's the most efficient way to sort": Video of Obama and Schmidt on YouTube, www.youtube.com/watch?v=k4RRi_ntQc8&feature=related.

63 "In general, we're not trying to fill a particular job": Setty interview, April 7, 2010.

64 woman who got the goods: Lorraine, "Google Cheat View." A blogger, "Idiot Forever," later claimed he submitted the story to *The Sun* as a hoax. See post at http://idiotforever.wordpress.com/2009/03/31/how-i-duped-the-sun/.

64 "It's a low bar for someone": Kaplan, "Want a Job at Google?"

64 "People are willing to tell you all sorts of stuff": Carlisle interview, April 7, 2010.

65 "provocative or inappropriate photographs": CareerBuilder, "Forty-Five Percent of Employers Use Social Networking Sites."

65 "I always try to get a list of people": Agrawal interview, June 8, 2010.

65 "It gets harder and harder": Carlisle interview, April 7, 2010.

66 "Not hiring someone for poor communication skills": Posted by "libation" on the *New York Times* site, as comment to Wortham, "More Employers Use Social Networks."

Chapter Five

68–69 alleged to have been devised by Steve Ballmer: Poundstone, *How Would You Move Mount Fuji?*, 79–80.

69 Feynman (a childhood hero of Sergey Brin's): Auletta, *Googled,* 28.

69 "We were offended at having four-digit numbers": Auletta, *Googled,* 32.

69 "A very senior Microsoft developer": See www.joelonsoftware.com/items/2005/10/17.html.

77 Tyma posed this question to his mother: See Tyma's blog post at http://paultyma.blogspot.com/2007/03/howto-pass-silicon-valley-software.html.

78 about twenty times faster than quicksort: With 1,000,000 records to sort, Mrs. Tyma's method requires 1,000,000 operations. Quicksort, and other optimal sorting algorithms, require on the order of 1,000,000 $\log_2$ (1,000,000) operations. Taking this at face value, Mrs. Tyma's method is approximately $\log_2$ (1,000,000), or 19.9+, times faster.

Chapter Six

80 "At Google, we believe in collaboration": Mohammad, blog post at http://allouh.wordpress.com/2009/04/14/interview-with-google/.

81 "You will have this 'lost in space feeling'": December 30, 2006, comment by "Daniel" on Shmula blog, www.shmula.com/31/my-interview-job-offer-from-google.

86 Gardner published a variant of this puzzle: Gardner, *Wheels, Life and Other Mathematical Amusements,* 30.

89 Gardner mentioned this one in his *Scientific American* column: Gardner, *The Scientific American Book of Mathematical Puzzles and Diversions,* 24, 28. See also Gardner, *The Unexpected Hanging and Other Mathematical Diversions,* 186, which gives the original year of publication as 1957.

92 Shriram... insisted on a blind test: Auletta, *Googled,* 43.

Chapter Seven

97 "Even if it's not a coding question": Carlisle interview, April 7, 2010.

Chapter Eight

106 "I'm going to ask you a few questions": Shontell interview, May 25, 2010.

110 confessed that he didn't know how much Google makes from Gmail: Shontell interview, May 25, 2010.

110 "The interview was going swimmingly": Orlowski, "Tales from the Google Interview Room."

111 "Let's guess that Gmail is one percent of the total revenue": One 2010 outside estimate had Gmail accounting for 0.3 percent of Google. See http://seekingalpha.com/article/196953-youtube-much-more-important-than-gmail-for-google.

112 "A job waiting on graduation day": According to the National Association of Colleges and Employers, less than a quarter of seniors had postgraduation job offers in April 2010. This was way down from 2007, when the figure was 52 percent. See Warner, "The Why-Worry Generation."

Chapter Nine

114 British media sensation: *Time,* "An Eggalitarian Education," 50. See also Gardner, *The Last Recreations,* 54.

123 "1. Drop egg from second floor": See December 6, 2006, post by "ptoner" on http://classic-puzzles.blogspot.com/2006/12/google-interview-puzzle-2-egg-problem.html.

Chapter Ten

125 "Unless we are certain of its weight": *Akron Beacon Journal,* "Head a Burger Standard."

125 Sonya "The Black Widow" Thomas won: *Akron Beacon Journal,* "Head a Burger Standard."

126 "How do they make M&M's?": I discuss this question in Poundstone, *How Would You Move Mount Fuji?,* 68, 159–60.

133 estimated and subtracted using standard formulas: See Brozek, "Densitometric Analysis of Body Composition."

134 "D'you know that the human head weighs 8 pounds?": "Mass of a Human Head" by Glenn Elert and students. http://hypertextbook.com/facts/2006/DmitriyGekhman.shtml.

134 "An adult human cadaver head": http://danny.oz.au/anthropology/notes/human-head-weight.html.

135 "The brain appears to be an efficient superhighway": Cohen, "Charting Creativity."

136 "might allow for the linkage of more disparate ideas": Cohen, "Charting Creativity."

136 "Hell, there are no rules here": See www.brainyquote.com/quotes/authors/t/thomas_a_edison.html.

136 "The goal is to find out where the candidates": BillR, November 19, 2009, post at http://blog.seattleinterviewcoach.com/2009/02/140-google-interview-questions.html.

Answers, Chapter Two

148 General Problem Solver...three cannibals and three missionaries: Newell and Simon, *Human Problem Solving.*

154 identified this puzzle's author as Frank Hawthorne: Gardner, *The Scientific American Book of Mathematical Puzzles and Diversions,* 33.

154 "have important applications in the cheese and sugarloaf industries": Quoted in Gardner, *The Scientific American Book of Mathematical Puzzles and Diversions,* 34. Putzer and Lowen's 1958 publication was a research memorandum issued by Convair Scientific Research Laboratory, San Diego.

155 Selvin argued that you should switch boxes: Selvin, "A Problem in Probability."

155 had to defend it in a follow-up letter: Selvin, "On the Monty Hall Problem."

155 "has been debated in the halls of the Central Intelligence Agency": Tierney, "Behind Monty Hall's Doors."

155 only 12 percent of those questioned: Granberg and Brown, "The Monty Hall Dilemma," 711.

156 "Certainly Monty Hall knows": Selvin, "A Problem in Probability."
158 "I wouldn't want to pick the other door": Granberg and Brown, "The Monty Hall Dilemma," 718.
158 "Even Nobel physicists systematically give": Vos Savant, *The Power of Logical Thinking,* 15.
161 Einstein equivalence principle says: You have to exclude subtler gravitation experiments involving tidal forces or such exotica as gravity waves and black holes.

Answers, Chapter Four

171 2004 *Emergency Evacuation Report Card:* Cox, "Emergency Evacuation Report Card."
172 school buses have greater capacity: See Cox, "Emergency Evacuation Report Card," note at bottom of p. 25.
172–73 Golden Gate Bridge has had reversible lanes: See www.goldengatebridge.org/research/facts.php#VehiclesCrossed.
179 "class Chicken": See other examples on "Ace the Interview," www.acetheinterview.com/questions/cats/index.php/fundamental/2007/09/17/chicken-by-spencer.
181 Want to convert between miles and kilometers?: This *is* a meaningless coincidence. The length of a mile in kilometers (1.609) just happens to approximate the ratio between adjacent Fibonacci numbers (about 1.618 when the numbers are large). See Peteris Krumins's 2010 post, "Using Fibonacci Numbers to Convert from Miles to Kilometers and Vice Versa," at www.catonmat.net/blog/using-fibonacci-numbers-to-convert-from-miles-to-kilometers.
185 "What could be more mystical": Crease, "The Greatest Equations Ever."
185 "Like a Shakespearean sonnet": Nahin, *Dr. Euler's Fabulous Formula,* 1.
186 "I know of scarcely anything": Galton, "President's Address," 495–96.
186 "It is more important to have beauty": Dirac, "The Evolution of the Physicist's Picture of Nature," 47.

Answers, Chapter Five

190 Primes Pages: primes.utm.edu.

Answers, Chapter Six

202 Conway proved some original and (semi)serious results: Conway, "The Weird and Wonderful Chemistry of Audioactive Decay." The sequence also figures in Clifford Pickover's 2001 book, *The Cuckoo's Egg.*

Answers, Chapter Eight

222 random packing occupies anywhere from 55 to 64 percent: Cartlidge, "The Secrets of Random Packing."

Answers, Chapter Nine

231 "Any one who considers arithmetical methods": Von Neumann, "Various Techniques Used in Connection with Random Digits."
242 Mayans were building them as early as 900 BC: Paterson, "Maximum Overhang," 1–2.
242 brick-stacking puzzle: Gardner, "Some Paradoxes and Puzzles Involving Infinite Series."

Answers, Chapter Ten

253 "looked like snot": http://blogs.chron.com/sciguy/archives/2006/02/ill_bet_you_did.html.
253 findings were published in a 2004 article: Gettelfinger and Cussler, "Will Humans Swim Faster or Slower in Syrup?"
254 flow of water is much less turbulent for a fish: Gettelfinger and Cussler, "Will Humans Swim Faster or Slower in Syrup?," 2647; Cussler interview, June 16, 2010.
254 computer science "is probably of no use": Cussler interview, June 16, 2010.
255 "The best swimmer": Hopkin, "Swimming in Syrup Is As Easy As Water."
255 "The reasons for this are complicated": See www.mitadmissions.org/topics/life/boston_cambridge/no_time_for_your_stupid_questi.shtml.

Postscript

256 Four questions traditionally posed: Numerous versions of this have been circulated in e-mails. I can't vouch for the e-mail claim that it was devised at Anderson Consulting (now Accenture) and that "around 90% of the professionals they tested got all questions wrong, but many preschoolers got several correct answers." The giraffe question is asked in job interviews, as a joke of course, at many companies besides Accenture.

Bibliography

Akron Beacon Journal. "Head a Burger Standard." January 18, 2006. www.redorbit.com/news/science/361388/new_coffee_shop_in_w_akron_plans_oodles_of_noodles/index.html.

Arango, Tim. "Present-Day Soapbox for Voices of the Past (with a Web Site)." *New York Times,* November 30, 2009.

Associated Press. "Study: Older Americans Staying Put in Jobs Longer." September 3, 2009.

Auletta, Ken. *Googled: The End of the World As We Know It.* New York: Penguin, 2009.

Beatty, Richard W., and Craig Eric Schneier. *Personnel Administration: An Experiential/Skill-Building Approach.* Reading, Mass.: Addison-Wesley, 1977.

Bernard, Tara Siegel. "Google to Add Pay to Cover a Tax for Same-Sex Benefits." *New York Times,* June 30, 2010.

Borelli, Giovanni Alfonso. *On the Movement of Animals.* Translated by P. Maquet. Berlin: Springer-Verlag, 1989.

Bray, Douglas W., Richard J. Campbell, and Donald L. Grant. *Formative Years in Business: A Long-Term AT&T Study of Managerial Lives.* New York: Wiley, 1974.

Brozek, Josef, Francisco Grande, Joseph T. Anderson, and Ancel Keys. "Densitometric Analysis of Body Composition: Revision of Some Quantitative Assumptions." *Annals of the New York Academy of Sciences* 110 (2006): 113–140.

Bryant, Adam. "On a Scale of 1 to 10, How Weird Are You?" *New York Times,* January 9, 2010.

CareerBuilder.com. "Forty-Five Percent of Employers Use Social Networking Sites to Research Job Candidate, CareerBuilder Survey Finds." Press release, August 19, 2009, www.careerbuilder.com/share/aboutus/pressreleasesdetail.aspx?id=pr519&sd=8/19/2009&ed=12/31/2009&siteid=cbpr&sc_cmp1=cb_pr519_&cbRecursionCnt=3&cbsid=22b26a56aa6241049185fc24b7298023-304332027-JP-5.

Cartlidge, Edwin. "The Secrets of Random Packing." *Physics World*, May 8, 2008.

Clifford, Stephanie. "Bug by Bug, Google Fixes a New Idea." *New York Times*, October 4, 2009.

Cohen, Patricia. "Charting Creativity: Signposts of a Hazy Territory." *New York Times*, May 7, 2010.

Conway, J. H. "The Weird and Wonderful Chemistry of Audioactive Decay." *Eureka* 46 (1986): 5–18.

Cox, Wendell. "Emergency Evacuation Report Card." American Highway Users Alliance, 2006. www.highways.org/pdfs/evacuation_report_card2006.pdf.

Crease, Robert P. "The Greatest Equations Ever." *Physics World*, October 6, 2004.

Cureton, Edward E. "Validity, Reliability and Baloney." *Educational and Psychological Measurement* 10 (1950): 94–96.

Dasgupta, Sanjoy, Christos Papadimitriou, and Umest Vazirani. *Algorithms*. New York: McGraw-Hill, 2008.

Dirac, Paul A. M. "The Evolution of the Physicist's Picture of Nature." *Scientific American*, May 1963.

Dunnette, Marvin D., and Bernard M. Bass. "Behavioral Scientists and Personnel Management." *Industrial Relations* 2 (1963): 115–30.

Edmans, Alex. "Does the Stock Market Fully Value Intangibles? Employee Satisfaction and Equity Prices," 2009. http://ssrn.com/abstract=985735.

Feynman, Richard, Robert B. Leighton, and Matthew Sands. *The Feynman Lectures on Physics*. Reading, Mass.: Addison-Wesley, 1963–65.

Frommer, Dan. "It's Harder to Get a Job at the Apple Store Than It Is to Get Into Harvard." *Yahoo! Finance*, November 12, 2009.

Galton, Francis. "President's Address." *The Journal of the Anthropological Institute of Great Britain and Ireland*, 15, 489–99, 1886.

Gamow, George, and Marvin Stern. *Puzzle-Math*. New York: Viking, 1958.

Gardner, Martin. *The Scientific American Book of Mathematical Puzzles and Diversions*. New York: Simon and Schuster, 1959.

———. *The Second Scientific American Book of Mathematical Puzzles and Diversions*. New York: Simon and Schuster, 1961.

———. "Some Paradoxes and Puzzles Involving Infinite Series and the Concept of Limit." *Scientific American*, November 1964, 126–33.

———. *Mathematical Carnival.* New York: Knopf, 1975.
———. *The Unexpected Hanging and Other Mathematical Diversions.* New York: Simon and Schuster, 1969.
———. *Wheels, Life and Other Mathematical Amusements.* New York: W. H. Freeman, 1983.
———. *The Last Recreations: Hydras, Eggs, and Other Mystifications.* New York: Copernicus, 1997.
Gettelfinger, Brian, and E. L. Cussler. "Will Humans Swim Faster or Slower in Syrup?" *American Institute of Chemical Engineers Journal* 50 (2004): 2646–47.
Glatter, Jonathan D. "Another Hurdle for the Jobless: Credit Inquiries." *New York Times,* August 6, 2009.
Goodman, Peter S. "U.S. Job Seekers Exceed Openings by Record Ratio." *New York Times,* September 26, 2009.
Granberg, Donald, and Thad A. Brown. "The Monty Hall Dilemma." *Personality and Social Psychology Bulletin* 21 (1995): 711–29.
Guilford, J. P. *Way Beyond the IQ.* Buffalo, N.Y.: Creative Education Foundation, 1977.
Gunter, Barrie, Adrian Furnham, and Russell Drakeley. *Biodata: Biographical Indicators of Business Performance.* London: Routledge, 1993.
Guynn, Jessica. "Tech Firms Try to Outperk One Another." *Los Angeles Times,* March 28, 2010.
Haldane, J. B. S. "On Being the Right Size." *Harper's Monthly* 152 (1926): 424–27.
Hansell, Saul. "Google Answer to Filling Jobs Is an Algorithm." *New York Times,* January 3, 2007.
Hart, Peter E., Nils J. Nilsson, and Bertram Raphael. "Correction to 'A Formal Basis for the Heuristic Determination of Minimum Cost Paths.'" *SIGART Newsletter* 37 (1972): 28–29.
Helft, Miguel. "An Auction That Google Was Content to Lose." *New York Times,* April 4, 2008.
———. "Google Makes a Case That It Isn't So Big." *New York Times,* June 28, 2009.
Hopkin, Michael. "Swimming in Syrup Is As Easy As Water." *NatureNews,* September 20, 2004. www.nature.com/news/2004/040920/full/news040920-2.html.
International Federation of Competitive Eating. "Sonya Thomas Retains Big Daddy Burger Title." Major League Eating, January 21, 2006. www.ifoce.com/news.php?action=detail&sn=361.
Isidore, Chris. "Out-of-Work Job Applicants Told Unemployed Need Not Apply." *CNNMoney,* June 16, 2010.
Iyer, Bala, and Thomas H. Davenport. "Reverse Engineering Google's Innovation Machine." *Harvard Business Review,* April 2008, 59–68.

Kahneman, Daniel. Nobel Prize Autobiography, 2002. http://nobelprize.org/nobel_prizes/economics/laureates/2002/kahneman-autobio.html.

Kaplan, Michael. "Want a Job at Google? Try These Brainteasers First." *Business 2.0,* August 30, 2007.

Laakmann, Gayle. "Cracking the Technical Interview," 2009. CareerCup.com.

LaBarbera, Michael C. "The Biology of B-Movie Monsters," 2003. http://fathom.lib.uchicago.edu/2/21701757.

Levering, Robert, and Milton Moskowitz. "What It Takes to Be #1: Genentech Tops the 2006 *Best Companies to Work For* in America List." Great Place to Work Institute, 2006. www.greatplacetowork.com.

Levering, Robert, Milton Moskowitz, and Michael Katz. *The 100 Best Companies to Work For in America.* Reading, Mass.: Addison-Wesley, 1984.

Lohr, Steve. "John W. Backus, 82, Fortran Developer Dies." *New York Times,* March 20, 2007.

Lorraine, Veronica. "Google Cheat View." *Sun* (London). March 31, 2009.

Lyons, Daniel. "The Customer Is Always Right." *Newsweek,* January 4, 2010.

Martin, Robert A. "Confessions of an Interviewer." *MBA,* January 1975.

McHugh, Josh. "Google vs. Evil." *Wired,* no. 11.01 (2003).

Merrell, Gerald P. "Have Your Google People Talk to My 'Googol' People." *Baltimore Sun,* May 16, 2004.

Millar, Garnet W. *E. Paul Torrance, "The Creativity Man."* Norwood, N.J.: Ablex Publishing, 1995.

Moggridge, Geoff. *How to Get into Oxford and Cambridge: Beating the Boffins.* Cambridge: PGR Publishing, 1998.

Nahin, Paul J. *Dr. Euler's Fabulous Formula: Cures Many Mathematical Ills.* Princeton, N.J.: Princeton University Press, 2006.

Neumann, John von. "Various Techniques Used in Connection with Random Digits." In *Monte Carlo Method,* edited by A. S. Householder, G. E. Forsythe, and H. H. Germond. Washington, D.C.: National Bureau of Standards, 1951.

Newell, Allen, and Herbert A. Simon. *Human Problem Solving.* Englewood Cliffs, N.J.: Prentice Hall, 1972.

Orlowski, Andrew. "Tales from the Google Interview Room." *Register,* January 5, 2007.

Patterson, Mike, Yuval Peres, Mikkel Thorup, Peter Winkler, and Uri Zwick. "Maximum Overhang," 2007. www.math.dartmouth.edu/~pw/papers/maxover.pdf.

Patterson, Scott. *The Quants: How a Small Band of Math Wizards Took Over Wall Street and Nearly Destroyed It.* New York: Crown, 2009.

Pegg, Ed, Jr., and Eric W. Weisstein. "Mathematica's Google Aptitude." *Mathworld,* October 13, 2004. http://mathworld.wolfram.com/news/2004-10-13/google/.

Phear, J. B. *Elementary Mechanics.* Cambridge: Macmillan, 1850.

Pickover, Clifford A. *Wonders of Numbers: Adventures in Mathematics, Mind, and Meaning.* Oxford: Oxford University Press, 2001.

Poundstone, William. *How Would You Move Mount Fuji? Microsoft's Cult of the Puzzle: How the World's Smartest Companies Select the Most Creative Thinkers.* New York: Little, Brown, 2003.

Selvin, Steve. "A Problem in Probability." Letter to the editor. *American Statistician* 29 (1975): 67.

———. "On the Monty Hall Problem." Letter to the editor. *American Statistician* 29 (1975): 134.

Stone, Dianna L., and Gwen E. Jones. "Perceived Fairness of Biodata as a Function of the Purpose of the Request for Information and Gender of the Applicant." *Journal of Business and Psychology* 11 (1997): 313–23.

Thaler, Richard. "Mental Accounting and Consumer Choice." *Marketing Science* 4 (1985): 199–214.

Tierney, John. "Behind Monty Hall's Doors: Puzzle, Debate, and Answer?" *New York Times,* July 21, 1991.

Time. "An Eggalitarian Education." May 18, 1970, 50.

Torrance, E. Paul. *Guiding Creative Talent.* Englewood Cliffs, N.J.: Prentice-Hall, 1962.

Tugend, Alina. "Getting Hired, Never a Picnic, Is Increasingly a Trial." *New York Times,* October 9, 2009.

Viereck, George Sylvester. "What Life Means to Einstein." *Saturday Evening Post,* October 26, 1929, 117.

Vogel, Steven. "Living in a Physical World: III. Getting Up to Speed." *Journal of Bioscience* 30 (2005): 303–12.

Vogelstein, Fred. "Search and Destroy." *Fortune,* May 2, 2005.

Vos Savant, Marilyn. *The Power of Logical Thinking.* New York: St. Martin's Press, 1996.

Warner, Judith. "The Why-Worry Generation." *New York Times,* May 24, 2010.

Weekley, Jeff. "Biodata: A Tried and True Means of Predicting Success." Accessed October 9, 2010. www.kenexa.com/ResourceCenter/ThoughtLeadership/Biodata-A-Tried-and-True-Means-of-Predicting-Succ.

Wortham, Jenna. "More Employers Use Social Networks to Check Out Applicants." *New York Times,* August 20, 2009.

Yen, Yi-Wyn. "YouTube looks for the Money Clip." *Fortune,* March 25, 2008.

Index

Note: Page numbers in parentheses indicate answers to questions.

About the Author

William Poundstone is the author of twelve books, including *How Would You Move Mount Fuji?* and *Fortune's Formula,* which was the Amazon Editors' Pick for the number one nonfiction book of the year. He has written for the *New York Times, Harper's, Harvard Business Review,* and the *Village Voice,* among other publications. He lives in Los Angeles.